SpringerBriefs in Molecular Science

History of Chemistry

Series Editor

Seth C. Rasmussen, Fargo, USA

For further volumes:
http://www.springer.com/series/10127

Mary Virginia Orna

The Chemical History
of Color

 Springer

Mary Virginia Orna
Department of Chemistry
The College of New Rochelle
New Rochelle NY
USA

ISSN 2212-991X
ISBN 978-3-642-32641-7   ISBN 978-3-642-32642-4   (eBook)
DOI 10.1007/978-3-642-32642-4
Springer Heidelberg New York Dordrecht London

Library of Congress Control Number: 2012948179

Printed on acid-free paper

Springer is part of Springer Science+Business Media (www.springer.com)

This volume is dedicated, from A to Z, to:

Alice Alexander (1903–1992), one of the most colorful persons who has graced my life

Mary Jane Robertshaw, who got the color wheels turning in the first place, and

Zvi C. Koren, who continues to color my world

Preface

Color has been an exciting and enjoyable part of human life ever since the color-sensitive eye evolved over a million years ago. However, the junction between color and chemistry, and color and history, is of more recent origin. The first recorded use of chemistry to manufacture a color is the stunning set of cave paintings found in the Grotte Chauvet in Southern France. Executed over 32,000 years ago (20,000 years earlier than Lascaux!), they are a testimony to early humans' ability to create beauty and to engage in abstract thinking. This volume traces the history of color usage as a chemical endeavor from the earliest records to the present day. It is a trajectory that is more or less direct since the three requisites of history, chemistry, and color function somewhat like a triple point in a phase diagram: they keep us on target! Nonetheless, the history of color chemistry finds stopovers in color physics, atomic theory, ancient dye production, medieval pigment synthesis, organic structural chemistry, and on up to the development of the modern chemical and pharmaceutical industries. It is a journey peppered with outstanding and fascinating personalities, and the difficulties they experienced in delving into the mysteries of color. Color, as we shall see, is not only a delight to the eye, but often a manifestation of the ultimate nature of the colored substance itself. Color pervades every aspect of our lives, our consciousness, our perceptions, our language, our useful appliances and tools, our playthings, our entertainment, our health, our diagnostic apparatus—all based in no small part on chemistry. As such, color is a universal experience and phenomenon. Its chemical history, as you shall see, is no less so.

Acknowledgments

I am indebted to so many persons who have been helpful in the development of this volume. In particular, I would like to thank Ashley Augustyniak, Alfred, Linda and Daniel Bader, Christina Blay, Antonia Clark, Robin J. H. Clark, Carmen Giunta, Jack Harrowfield, Ernst Hempelmann, Richard Hermens, Donna Jenkins, Frederick H. Kasten, Jan Kochansky, Zvi C. Koren, Kathleen Mannino, Harold T. McKone, Daniel Rabinovich, Roger Rea, Carolyn Reid, Silvia Rozenberg, Adrian-Mihail Stadler, and Anthony S. Travis. I would also like to thank the Chemical Heritage Foundation for the use of many images from its picture collection. I am very grateful for the comments and suggestions of my editor, Seth Rasmussen. Special thanks are due to Marco Fontani for his invaluable help in obtaining hard-to-get references, and finally, I would like to acknowledge with great thanks Philip Ogata for his meticulous reading of the manuscript draft, for his many helpful suggestions, and for his encouragement all along the way.

I would also like to acknowledge the invaluable reference help I received at the Library of the College of New Rochelle, at the Donald F. and Mildred Topp Othmer Library, Chemical Heritage Foundation, and at the Library of the Academy of Natural Sciences, Philadelphia, Pennsylvania.

Contents

Chapter 1
Introduction: Colors, Natural and Synthetic, in the Ancient World

1.1 Introduction

Color is important. It gives life to everything people do, think, and even say. In the days when color television sets were expensive, and the corner bar was in sole possession of one, the neighborhood center became the corner bar. Color sensation is a universal human experience. From the beginning of recorded history, references to color abound in connection with every aspect of human life. Color is an integral part of poetry, descriptive narratives, and technical treatises. It is a phenomenon that everyone appreciates; a "colorless" world composed only of blacks, whites, and grays, like that old television set, is difficult to imagine today. Color is also a property of materials that has been an integral part of human experience in every age and civilization. It has caused humans to wonder about its origin [1] and to experiment in its production. In fact, manipulation of colored pigments is perhaps one of the oldest applications of what is now called chemistry.

Color is a subject that sprawls across the four enormous disciplines of physics, chemistry, physiology and psychology. Any attempt to arrive at a satisfactory definition of the word involves a dip into each of these disciplines and into the areas that overlap them. Furthermore, the word color can be applied to not only the visual arts, but also to the worlds of law, music, dance, verbal expression, and personality traits. The *American Heritage Dictionary* [2] defines color in eighteen different ways, and devotes a half-page article to explaining the definitions of color names used elsewhere in the dictionary.

Through its very title, this work will narrow down this vast subject to chemistry, and specifically to chemistry as seen through the lens of history, although it will be necessary from time to time to allude to color's profound relationship to the other disciplines named above. The impact of light and color on society, scientific theory, and human self-understanding will be explored by fleshing out the history of color from the chemical point of view, beginning with the first recorded uses of color down to the development of our modern chemical industry, looking at some theoretical physics along the way. We will see that color pervades every aspect of

M. V. Orna, *The Chemical History of Color*, SpringerBriefs in History of Chemistry, DOI: 10.1007/978-3-642-32642-4_1, © The Author(s) 2013

our lives, our consciousness, our perceptions, our useful appliances and tools, our playthings, our entertainment, our health, our diagnostic apparatus—all based in no small part on chemistry.

1.2 Working Definition and Nature of Color

Although the ordinary observer takes color largely for granted, it is a very complex phenomenon involving the physics of a light source, the chemistry of the colored object which modifies the light falling upon it, and the psychophysical response of the observer to the resulting stimulus. There can be no color, therefore, without light. It is light that impinges on and is reflected or that is emitted from an object that allows it to be seen. Objects that emit light, like the sun, the stars, and a neon light, are called luminous bodies; objects which light impinges, like the moon and the cover of a book, are called illuminated bodies. For our purposes, color may be defined as that aspect or perception of a thing that is caused by the different qualities of the light reflected, scattered, absorbed, transmitted, refracted, or emitted by it. Thus we see that the production of color depends on three factors: the light source, the object being illuminated, and the receiver, most often a human eye, that perceives the color [3]. More detail regarding these aspects will be examined in Chap. 2.

1.3 Natural and Synthetic Colors in the Ancient World

The variety and richness of colored materials used to produce artifacts has grown tremendously since the artists who decorated the Grotte Chauvet in southern France set to work on their *opus magnum* about 32,000 years ago, an astounding result of accelerator mass spectrometry measurements carried out on the actual pigments themselves [4]. The natural colors they employed were a variety of shades of ochres, or iron earths—red, yellow, brown, or golden—depending upon the degree of hydration of the key compound, iron(III) oxide. The actual coloring material was a minor constituent of the clay earth, at most 20 %, and the best varieties were found in deposits in southern France. Another natural color found in Grotte Chauvet was black manganese dioxide. The second black color was actually synthetic: charcoal prepared by burning hard woods in a limited supply of air [5]. Figure 1.1 shows several of these shades in the cave-artist's depiction of a steppe bison as commemorated on a Romanian postage stamp.

In the intervening tens of thousands of years between the earliest cave paintings and recorded history, the artist's palette expanded to include many other natural and synthetic colors in a variety of shades and hues. Stone plates for grinding and mixing face powders and eye painting unearthed from Egyptian tombs date back to about 6000 years BCE. Egyptian blue (calcium copper tetrasilicate, or

Fig. 1.1 Paleolithic artists worked their chemistry on the walls of Grotte Chauvet 32,000 years ago. This Romanian stamp reproduces the splendid earthen tones in the image of a steppe bison

cuprorivaite, $CaCuSi_4O_{10}$) and vermilion (mercury(II) sulfide, HgS) were among the earliest of the manufactured pigments. In addition to these colorants and the naturally occurring iron and manganese earth pigments mentioned earlier, the Egyptians, by 3000 BCE, had succeeded in expanding the palette to include red lead (Pb_3O_4), malachite (basic copper carbonate), orpiment (arsenic trisulfide), and red madder (from the roots of the perennial *Rubia tinctorum*). As time went on, other pigments were added and by 1300, the list stood at about three dozen pigments in common use [6]. Some naturally occurring coloring matter was also derived from plants and animals, although their use by artists has been limited due to their instability. Two of the most famous of these colorants are indigo and carmine-cochineal. Table 1.1 lists the most commonly used of these colorants, although many more will be introduced in subsequent chapters.

1.3.1 Blue Coloring Materials

Some of the coloring agents listed above deserve additional mention. Among these is the material commonly called Egyptian Blue. This material was not only the first synthetic pigment (if we exclude charcoal), but it was one of the first materials from antiquity to be examined by modern scientific methods. Egyptian Blue, as is evident from its name, first occurred in Egypt during the 3rd millennium BCE, and during the subsequent 3,000 years, its use as a pigment and a coloring material for decorative objects spread throughout the Middle East and to the farthest limits of the Roman Empire. A sample of Egyptian Blue was found in 1814 during the Pompeii excavation, and captured the interest of notable scientists such as Sir Humphry Davy. It was not until 1884 that the chemical composition of this glass-like colorant was determined by F. Fouqué [7], although the actual material, being quite heterogeneous, was found to differ in many characteristics such as color, particle size, and hardness depending upon where it was produced. Modern

Table 1.1 Common coloring materials used from early times (before 1300 CE)

Common name	Chemical identity	Starting date	Comments
Azurite (blue)	Basic copper carbonate, $2CuCO_3Cu(OH)_2$	From prehistoric times	Most important in wall paintings in the East; known in ancient Egypt
Carmine (Cochineal carmine and kermes carmine)	Cochineal: carminic acid, a glucopyranose derivative of alizarin; kermes: kermesic acid, a poly-substituted derivative of alizarin	Kermes— from antiquity Cochineal— Pre-conquest	Kermes—one of the oldest organic colors known; Cochineal -Traditional red color of pre-Hispanic Mexico
Charcoal (black)	Elemental carbon	From prehistoric times	Also called carbon black; produced by dry distillation of wood in a closed vessel
Cinnabar (Vermilion, red)	Mercury(II) sulfide, HgS	From antiquity	One of the oldest known synthetic pigments
Egyptian blue	Calcium copper tetrasilicate, $CaCuSi_4O_{10}$	IV Egyptian dynasty or before	Crystalline compound containing some glass impurity
Gamboge (mustard yellow)	Gum resin from the tree Genus *Garcinia*	From antiquity	Indigenous to India, Ceylon, Thailand
Green earth	Green clay containing celadonite, a mica-group mineral	From pre-classical times	Used in ancient Roman wall paintings
Indigo (blue)	Indigotin, $C_{16}H_{10}N_2O_2$	One of the oldest coloring materials known	Used in painting Roman war shields
Iron earths, ochres (red, yellow, brown)	Iron(III) oxide, Fe_2O_3; $Fe_2O_3nH_2O$	From prehistoric times	Found in prehistoric cave paintings
Lamp black	Nearly pure amorphous C	From antiquity	Preparation described by Pliny the Elder
Lead white	Basic lead carbonate, $2PbCO_3Pb(OH)_2$	From antiquity	Preparation described by Pliny the Elder
Madder (red)	Chiefly alizarin, $C_{14}H_8O_4$	From classical times	Natural (Root of *Rubia tinctorum*)
Malachite (green)	Basic copper carbonate, $CuCO_3Cu(OH)_2$	From prehistoric times	Oldest known bright green pigment
Orpiment (yellow)	Arsenic trisulfide, As_2S_3	From prehistoric times	Derives its name from a corruption of *auripigmentum*, i.e., gold color

(continued)

Table 1.1 (continued)

Common name	Chemical identity	Starting date	Comments
Realgar	Orange-red arsenic sulfide, AsS or As_4S_4	From antiquity	Confused for red lead by ancients
Saffron (yellow)	Complex mixture of 150 compounds	From prehistoric times	From dried stigmas of *Crocus sativus*
Ultramarine blue (Natural)	Blue clathrate compound derived from *lapis lazuli*	From ancient times	Very costly in medieval times
Verdigris (green)	Dibasic acetate of copper, $Cu(C_2H_3O_2)_2 2Cu(OH)_2$	From antiquity	Other copper compounds, including carbonates, are also called verdigris
Weld (yellow)	Flavone, $C_{15}H_{10}O_2$; 2-phenylchromone	From antiquity	Extracted from parts of *Reseda luteola*

laboratory reproduction of Egyptian Blue indicates a probable single firing and a two-stage firing cycle in the production of the ancient material [8].

Several other blue coloring materials are of great interest in this context. Azurite occurs naturally as a crystalline material throughout Europe and the countries of the former Soviet Union. Its use may date from the fourth dynasty in Egypt, but the availability of Egyptian Blue in antiquity made its widespread employment unnecessary. Azurite was by far the most important blue pigment used by European painters throughout the Early and High Middle Ages. It can be prepared synthetically by precipitating copper(II) hydroxide from a solution of copper(II) chloride with lime (CaO), and then digesting the resulting precipitate with potassium carbonate and lime, although there is strong evidence that side reactions producing basic copper(II) chlorides may also take place [9]. The resulting synthetic product was called blue verditer to distinguish it from its naturally occurring counterpart.

Copper-based blue materials manufactured from recipes found in medieval artists' manuals present a very complicated chemical profile. Long before the synthesis of blue verditer became commonplace, there is strong evidence that the methodology for its production evolved over a period of at least ten centuries, beginning with starting materials such as elemental copper and even elemental silver, which the early synthesizers thought was "very pure". By examining these recipes, one can discern a gradual realization that the oxidized form of copper, i.e., copper contained in one of its compounds such as the acetate or the chloride, would produce faster and more homogeneous results [10, 11]. Although pure compounds such as tetra-μ-acetatobisdiaquocopper(II) [11, 12] and calcium copper hexahydrate [13, 14] were produced, other compounds and mixtures of compounds defied characterization. Some pure compounds have been identified that were formed when acetic acid vapor, water vapor and air were allowed to act on copper and copper alloys, and some forms of azurite have been identified by Gettens and Stout [15], but none of these compounds resemble the copper-based blue pigments produced from the recipes cited above. It is well-known that copper and copper-silver alloys yield a

Fig. 1.2 Manuscript
production in a monastery.
© 1997, Martha Counihan

variety of complex compounds depending upon the materials that react with them, and future research may reveal some new exotic compounds. Some of these compounds are definitely green, but some involving reactions of copper with such reagents as sour milk and cream of tartar can form some blue copper compounds which include lactates and tartrates. These, in addition to the acetates and carbonates discussed here, indicate the degree of sophistication attained by medieval artists and craftsmen long before the advent of the modern chemical theory that would provide a theoretical basis for these syntheses [16].

The use of another blue pigment, natural ultramarine, derived from the semi-precious mineral lapis lazuli, can be traced to sixth and seventh century wall paintings in Afghanistan. It was introduced as a rare import to European artists in the thirteenth century following the spectacular journeys of Marco Polo. During the subsequent three centuries, good quality ultramarine was as costly as gold, and patrons usually agreed to supply the pigment for a commissioned work of art or to pay for it separately at market price [9, 17]. It was a staple pigment for manuscript production throughout the Middle Ages, as the cartoonist who produced Fig. 1.2 assumed.

The only natural blue organic colorants in antiquity, often used as dyes, were the indigo and woad plants, both containing the identical coloring matter, indigo, or indigotin. It was one of the oldest coloring matters known. The chemical identity of this colorant was, of course, unknown until the advent of modern chemistry, and both materials were thought to be distinctly different from one another. Although woad has always been associated with its vegetable origin, it was once thought, at least in England, that indigo was of mineral origin. The indigo-bearing plant, *Indigofera tinctoria*, was formerly grown all over the world, but the synthetic product has replaced the natural product since 1900. Woad is obtained from *Isatis tinctoria*, a herbaceous biennial indigenous to Europe and the Middle East [15], and contains as little as 1/30 the amount of indigo in its coloring matter. Indigo is, to this day, one of the few naturally occurring dyes in wide use [18], most of the time as its synthetic counterpart.

1.3.2 Red Coloring Materials

One of the most important red colors from ancient times is mercury(II) sulfide, HgS, known to the artist as vermilion or as the ore, cinnabar. The naturally occurring mineral is widely, but not abundantly, distributed throughout Europe, the Middle East, the Far East, and the former Soviet Union The famous mines of Almaden, in Spain, remain to this day the most important source of cinnabar in the world. The ore needed to be broken into small pieces, ground finely, and then repeatedly washed and heated until reasonably pure. This process was laborious and time consuming, so much so that early on, recipes appeared for manufacturing vermilion from its elements. It appears that the earliest recipes came from China and entered western literature around the 8th century in a Latin manuscript of Greek origin known as the Lucca manuscript (No. 490), or more completely as *Compositiones ad Tingenda*. Synthesis by combining the mercury and sulfur with heating invariably led to a black material called "ethiops mineral" on account of its color. It was necessary to treat this compound, which was later realized to be what was called the β-form of mercury(II) sulfide, by heating and condensation (dry process) or by washing with a solution of soluble sulfides (wet process) to obtain the red α-form of the compound. Vermilion has been found on oracle bone inscriptions from China that date to 2000 BCE, and on Roman wall paintings as early as the first century CE [19].

Carmine is a red coloring material that has been used from ancient times in both hemispheres. In the Old World, the source of the red color was the kermes female scale insect (*Kermes vermilio*), a parasite that lived on chiefly the kermes or scarlet oak (*Quercus coccifera*). It has been known at least since the days of Moses; the color is mentioned in Exodus 26:1 in the directions for making the textile furnishings for the Temple. Color-makers in ancient Greece and Rome used it widely. Since it was abundant on the Iberian Peninsula, the Spaniards paid half their tribute to Rome with kermes grains. In the New World, the source of the red color was the wingless female scale insect *Dactylopius coccus*, indigenous to Mexico, Central and South America, and found chiefly on two host plants, the prickly pear and torch thistle cacti. It was in use as a textile dye from at least the old Peruvian Paracas culture that dates back to 700 BCE. Figure 1.3 is a textile made by the Paracas people of what is now coastal southern Peru. The pre-Columbian cultures of the Andes made exquisite textiles, which often depicted mythical stories and

Fig. 1.3 Fringed cape dyed with carmine cochineal from Paracas necropolis, Peru—Geometric cats embroidered on plain weave wool. Photograph by M.V. Orna

Fig. 1.4 Bodies of the female scale insect *Dactylopius coccus* from which the vast varieties of carmine-cochineal red are made. The dye is found in the egg sacs of the insect. Photograph by M.V. Orna

were sometimes used as markers of status by their owners. This piece from the Paracas Necropolis (south coast of Peru) dates from 0 to 100 CE. The red color is carmine-cochineal. Figure 1.4 is a photograph of the bodies of the female *Dactylopius coccus* from which the dye is made. Neither of these colorants is used much anymore in artistic endeavors because of their fugitive nature and the availability of more lightfast substitutes. They are used nowadays in cosmetics and in medical studies [20].

Another important organic red color that has served since ancient times for dyeing textile materials is madder. This colorant is derived from the roots of *Rubia tinctorum* and other plants in the *Rubiaceae* family; it consists of a mixture of alizarin, purpurin, and pseudopurpurin, and thus one might expect that the natural product would yield a variety of shades. It was described by Strabo, Dioscorides, Pliny the Elder, and in the Talmud. It has been identified on Egyptian textiles as early as 1300 BCE and in six paint pots of Greco-Roman origin that are now in the British Museum. In addition, textiles dyed with madder were sold at the St. Denis market, near Paris, during the seventh century CE, and Charlemagne himself promoted its cultivation. Numerous madder treatments were carried out over the centuries for brightening the color and enriching and refining the coloring material. A variant on madder red is used to this day for the French military colors [21].

1.3.3 Yellow Coloring Materials

Two closely related colorants, orpiment and realgar, have been in use from pre-historic times, having been found on wall paintings in Giza (Egypt) as early as 4000 BCE. Orpiment is a bright yellow colorant with the formula As_2S_3. The ancient Greeks called it *arsenikon* from the Greek word *arsenikos* meaning "male;" the Greeks believed that metals were of different sexes. Orpiment is found in ore veins throughout Europe as well as in the Middle East, China, and Japan. It also occurs in hydrothermal vents, as a hot-spring deposit, and as a volcanic sublimation product. Its chemical counterpart, realgar, AsS (or As_2S_2 or As_4S_4) has an orange-red color; its name comes from the Arabic *rahj al ghar*, powder of the mine. Realgar occurs as a minor constituent of certain ore veins associated with orpiment and other arsenic minerals. The ancients were well acquainted with the toxicity of both of these minerals [22].

The most important yellow dye in ancient and medieval times was weld, a flavone derivative extracted from the seeds, stems, and leaves of *Reseda luteola*, commonly known as dyer's rocket. Weld is resistant to atmospheric oxidation, rendering it quite lightfast and hence extremely popular and useful. In combination with the blue dye woad it was used to produce the Lincoln green made famous by Robin Hood and his merry men. Quercitron, a flavonol derivative, is much more susceptible to degradation by light and hence was not as important [23].

From this brief introduction to the natural and synthetic colors available in the ancient world, it is evident that color is an area where chemistry, physics, history, geography, mineralogy, geology, cultural anthropology, art, biology, psychology and a host of other areas of human endeavor intersect. In the next chapter, a discussion of the scientific basis of color will allow more precision regarding the kinds of substances classified as colorants and how they interact with light to modify it.

References

1. Gerritsen F (1974) Theory and practice of color. Van Nostrand-Reinhold, New York, pp 9–18
2. The American heritage dictionary (1985) 2nd college edition. Houghton Mifflin, Boston, p 293
3. Orna MV (1978) The chemical origins of color. J Chem Educ 55:478–484
4. Valladas H et al (2001) Radiocarbon AMS dates for Paleolithic cave paintings. Radiocarbon 43(2B):977–986
5. Curtis G (2006) The cave painters: probing the mysteries of the world's first artists. Knopf, New York, pp 18, 64

6. Orna MV, Goodstein M (1998) Chemistry and artists' colors. Spaulding, Wallingford 283
7. Fouqué F (1884) Bull Soc de Mines de France 12:36–37
8. Tite MS, Bimson M, Cowell MR (1984) Technological examination of Egyptian blue. In: Lambert JB (ed) Archaeological chemistry III. American Chemical Society, Washington, pp 215–242
9. Kühn H (1973) Terminal dates for paintings from pigment analysis. In: Young WJ (ed) Application of science in examination of works of art. Museum of Fine Arts, Boston, pp 199–205
10. Orna MV, Low MJD, Baer NS (1980) Synthetic blue pigments: ninth to sixteenth centuries. I. Literature. Stud Conserv 25:53–63
11. Orna MV, Low MJD, Julian MM (1985) Synthetic blue pigments: ninth to sixteenth centuries. II. 'Silver blue'. Stud Conserv 30:155–160
12. Brown GM, Chidambaram R (1973) Dinuclear copper(II) acetate monohydrate: a redetermination of the structure by neutron diffraction analysis. Acta Crystallogr B29:2393–2403
13. Smith CS, Hawthorne JG (1974) Trans Am Philos Soc 64(4):3–128
14. Langs DA, Hare CR (1967) J Chem Soc, Chem Commun 890–891
15. Gettens RJ, Stout GL (1966) Painting materials: a short encyclopedia. Dover, Garden City
16. Orna MV (1996) Copper-based synthetic medieval blue pigments. In: Orna MV (ed) Archaeological chemistry: organic, inorganic and biochemical analysis. American Chemical Society, Washington, pp 107–115
17. Plesters J (1966) Ultramarine blue: natural and artificial. Stud Conserv 11:62–91
18. Orna MV, Kozlowski AW, Baskinger A, Adams T (1994) Coordination chemistry of pigments and dyes of historical interest. In: Kauffman GB (ed) Coordination chemistry: a century of progress. American Chemical Society, Washington, pp 165–176
19. Gettens RJ, Feller RL, Chase WT (1993) Vermilion and cinnabar. In: Roy A (ed) Artists' Pigments: a handbook of their history and characteristics, vol 2. Oxford University Press, New York, pp 159–182
20. Schweppe H, Roosen-Runge H (1986) Carmine—cochineal carmine and kermes carmine. In: Feller RL (ed) Artists' pigments: a handbook of their history and characteristics, vol 1. National Gallery of Art, Washington, pp 255–283
21. Schweppe H, Winter J (1997) Madder and alizarin. In: FitzHugh EW (ed) Artists' Pigments: a handbook of their history and characteristics, vol 3. Oxford University Press, New York, pp 109–142
22. FitzHugh EW (1997) Orpiment and realgar. In: FitzHugh EW (ed) Artists' Pigments: a handbook of their history and characteristics, vol 3. Oxford University Press, New York, pp 47–79
23. Gregory PF, Gordon P (1983) Organic chemistry in colour. Springer-Verlag, New York

Chapter 2
Discovery of the Physics of Color

Anyone who has shared the company of three-year old children for an afternoon will recognize the conversation. Everything you say is challenged immediately with their favorite word—why? Newly arrived in the world and fascinated by all they see, three-year olds want to know the why of everything. They live in a world of mystery, wonder, and possibility—where magic is real and reality is magic. As we explore the discovery of the physics of color origin and perception, put yourself in the place of the three-year old who always wants to know why. Or put yourself in the place of the discoverers themselves who must have been filled with awe at what they learned and communicated to the world for the first time.

2.1 Theories on the Nature of Light

An inquisitive three-year old could indeed be a model for some of the scientists who people this chapter. First, there is Robert Hooke (1635–1703), described in one of his biographies as a "restless genius." Hooke was an indispensable assistant to Robert Boyle (1627–1691), having designed and built the air pumps so crucial to Boyle's development of his laws on gases. However, Hooke was an inventor and theoretician in his own right, being inquisitive on a number of problems that led to formulations of laws (Hooke's Law), inventions (microscope), paleonto-logical theories (he was the first person to realize the meaning and significance of fossils) and theoretical speculations (gravity). Among these speculations, and developed further by the ideas and work of Christiaan Huygens (1629–1695), was Hooke's proposal in 1665 on the nature of light: that light has the characteristics of a wave in an invisible medium permeating all space, solids, liquids, gases and vacuum. This medium they called the 'ether.' Huygens was able to show mathematically that the fundamental geometric laws of optics could be explained by assuming that a prism or lens slowed the speed of a light wave [1, 2].

Vigorously opposed to Hooke's and Huygens' wave theory was an intractable Isaac Newton (1643–1727) who, in his 1704 book, *Opticks or a Treatise on the*

M. V. Orna, *The Chemical History of Color*, SpringerBriefs in History of Chemistry, DOI: 10.1007/978-3-642-32642-4_2, © The Author(s) 2013

Fig. 2.1 Isaac Newton
commemorative stamp issued
by Germany in 1993

Reflexions, Refractions, Inflexions and Colours of Light [3] propounded his opinion that light consisted of a flux of imponderable particles based upon his observation of diffraction around a needle and bright bands of colored fringes on thin layers now known as Newton's ring. To explain these colors, Newton said light was a particle that had "fits." Huygens actually offered a more convincing explanation for the phenomena of reflection, refraction, and diffraction based upon wave nature. He suggested that the various colors travel at different velocities in different media, and therefore, each color has a different angle of refraction. Unfortunately, Huygens' theory was overshadowed by Newton's influence in Britain and Newton's idea that eventually came to be known as the particle theory of light dominated scientific thought for the succeeding 200 years.

Theories regarding the nature of light and color go back to the ancient Greeks. Aristotle is credited with making the first important contribution to what is now the modern theory of selective absorption [4]. It was Seneca, a Roman philosopher of the first century CE, who first noted that a prism reproduces the colors of the rainbow. Leonardo da Vinci (1452–1519) noticed that when light struck a water glass placed on a tabletop that it "spread out" as a colored image on the floor, but it remained for Isaac Newton in 1666 to formulate modern color theory on the basis of experiment. After taking his BA degree from Cambridge in 1665, Newton returned to his home at Woolsthorpe near Grantham because the threat of the Great Plague closed down the university. Starting at the age of 22, in a matter of 2 years, he had one of the most remarkable periods of creativity of any person who ever lived. One of these creative ideas that he developed became known as the *experimentum crucis*, the theory of colors based upon experiment. Figure 2.1 commemorates this event as well as the 350th anniversary of Newton's birth.

2.2 Newton's Famous Prismatic Dispersion Experiments

Newton allowed a small beam of sunlight (white, or achromatic, light) to pass through a prism in a darkened room. The light that emerged was no longer white

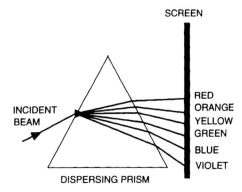

Fig. 2.2 A schematic diagram of Isaac Newton's famous prismatic dispersion experiment. © 1998, M.V. Orna

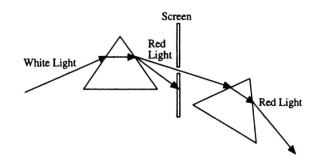

Fig. 2.3 A summary diagram of Newton's *"Experimentum crucis"* © 1998, M.V. Orna

light but exhibited a series of colors ranging from red, through orange, yellow, green, blue to violet (the visible spectrum; see Fig. 2.2).

He then asked if these colors, in turn, were made up of mixtures of other colors. To find out, he allowed his spectrum to fall upon a screen with hole in it. He then turned his prism until only red light passed through the hole. In back of the screen he placed a second prism and allowed the red light to pass through it. Newton reasoned that if red were a mixture of other colors, the second prism would disperse, or fan out, these colors in the same way that the first prism had dispersed the white light into its constituent colors. No such dispersion into additional colors occurred, nor were any of the other colors dispersed when they were tried in turn. Each color appeared to be the same although, as expected, each was dispersed to a greater degree by the second prism.

A summary of this important experiment is illustrated in Fig. 2.3. Newton concluded that red, orange, yellow, green, blue, violet were the fundamental colors recognized by the human eye and were not themselves mixtures of other colors. (Spectral colors are usually given in order starting with red, perhaps because red light is bent least by a prism.) Newton called this two-prism experiment his "Experimentum crucis." By it he conclusively demonstrated that sunlight is a mixture of six (seven, if one includes indigo, as did Newton) colors, that these colors could not be further dispersed into additional colors, and must therefore constitute the fundamental colors. In his own words, he concluded that "Light itself is a heterogeneous mixture of differently refrangible rays." Newton summarized the results as:

1. Sunlight consists of a mixture of all the colors observed in the prismatic spectrum (a word that Newton coined).
2. The prism is capable of dispersing the white light into its constituent colors. (Therefore, color was a property of the light not of the prism).

2.3 Consequences of Newton's Experiment

If Newton's first conclusion, that sunlight consists of a mixture of all the observed colors dispersed by the prism, then the dispersed colors should also be re-combinable into white light. To clinch this conclusion, therefore, Newton performed his recombination experiment. He placed a second, but inverted, prism in the path of the dispersed light and they did indeed, as predicted, recombine.

The observed variation of angle of refraction with color is due directly to the wave nature of the incident light, though Newton would have vigorously disputed this statement in his time. A series of experiments and theoretical development over a period of about 70 years, carried on by three great English scientists, gradually dismantled the one-sided Newtonian idea that light was corpuscular.

At the beginning of the nineteenth century, Thomas Young (1773–1829) obtained experimental evidence for the principle of interference by passing light through extremely narrow openings and observing the interference patterns. This experiment, along with the newly discovered polarized light by the Frenchman, Etienne Malus (1775–1812), allowed Young to conclude light had a transverse component, or in other words, his experiments could not be interpreted unless light was understood to have wavelike character.

About 165 years after Newton's experiment, in 1831, Michael Faraday (1791–1867) showed that charged particles produced magnetic fields and that both charges and magnets exerted an influence across space. Then in 1865, James Clerk Maxwell (1831–1879) theorized that light was energy of a special form: a wave in a magnetic and electrical field. He derived equations which described the propagation of light as a wave with an electric field perpendicular to a magnetic field and with both fields perpendicular to the direction of travel. Thus, the idea that light was electromagnetic radiation was conceived. Also, the idea that light was produced by an accelerated charged particle was developed. Light can produce colors by acting on charges in matter. Thus, vibrating ions and moving electrons can interact with light to modify the light.

A characteristic property of all electromagnetic radiation is the frequency of the field of oscillation, v, which remains invariant as the wave travels through any medium. The frequency is related to the velocity of the wave, c, and the wavelength, λ, by the equation $v\lambda = c$. It follows from this relationship that both λ and c

Table 2.1 The fundamental colors of the visible spectrum

Color	Wavelength range (nm)[a]	Band width (nm)	Frequency (s^{-1})[b]	Energy (eV)
Red	647.0–700.0	53	4.634–4.283	1.92–1.77
Orange	585.0–647.0	62	5.125–4.634	2.12–1.92
Yellow	575.0–585.0	10	5.214–5.125	2.16–2.12
Green	491.0–575.0	84	6.103–5.214	2.53–2.16
Blue	424.0–491.0	67	7.071–6.103	2.93–2.53
Violet	400.0–424.0	24	7.495–7.071	3.10–2.93

[a] A nanometer, nm, is one-billionth of a meter, a very small unit of length
[b] These values must be multiplied by 10^{14}. The unit s^{-1} is called a reciprocal second; it is essentially 1/s, or time divided into 1. This unit also represents the number of cycles per second in a light wave (cps), and is also known as hertz (hz)

must vary as a wave of a given frequency travels through different media since the frequency remains invariant.

As we examine a series of waves of given frequencies, we see that when the frequency is great, so is the energy of the wave. The equation governing this relationship was worked out by Albert Einstein (1879–1955) and Max Planck (1858–1947) and is thus called the Einstein–Planck relationship, $E = h\nu$, where h is the Planck constant with units of energy times time. A convenient value for h is 4.136×10^{-15} eV s. An electron volt (eV) is defined as the energy an electron gains when moved through a potential of one volt. If, for example, each electron "stored" in an ordinary 12-V automobile battery has a potential of 12 eV, then this amount of energy is expended by each electron as the battery discharges in use. The energies of electromagnetic radiation vary from more than 3×10^6 eV to less than 10^{-5} eV. The visible portion of this spectrum, i.e., the energy response range of the human eye, occupies only the very small region between about 1.7 and 3.1 eV. Each color in the visible spectrum has associated with it a range of frequencies from which can be calculated corresponding wavelength ranges. The fundamental colors of visible light along with their properties can best be summarized in tabular form as in Table 2.1.

2.4 The Visible Spectrum Examined

Table 2.1 contains a wealth of information and it would be well to examine it in detail. First of all, one notices that just a very small change in wavelength effects a change in perceived color. Secondly, each color does not consist of a single wavelength, but of a wavelength range. Although the visible spectrum covers the 400–700 nm wavelength range, all of the waves from 400 to 424 nm are seen as violet, all of the waves from 491 to 575 nm are seen as green, and so forth. These individual ranges are called pure spectral colors. It is difficult for the eye to distinguish between waves with wavelengths of 495 and 560 nm. Both waves are perceived as green. The wavelength range accounts for the "smear" effect seen when white light, or better, achromatic light, is dispersed by a prism. (It should also be noted that the dividing line between spectral colors is difficult to discern; their subdivision into six broad regions is somewhat arbitrary).

A third feature gleaned from the table is that some colors have broader wavelength ranges than others. The term "bandwidth" is used to denote the wavelength range of a given color. For example, the wavelength range of green light is 8.4 times greater than that of yellow light. This fact is also evident when achromatic light is dispersed by a prism. However, if one were to measure the widths of the green and yellow bands with a ruler, it would be found that the green band is not 8.4 times the width of the yellow band. This is because a prism does not disperse each wavelength of light to the same extent.

Another important feature gleaned from this table is the energy values of visible light. The last column gives the energies associated with each color of light. Red light has an energy range of 1.77–1.92 eV, whereas violet light has an energy range of 2.93–3.10 eV. Violet light is considerably more energetic than red light.

The energy differences between colors are exceedingly small because electron volts themselves are very small units. This shows what a very sensitive instrument the eye is. To distinguish between red and orange requires a remarkable ability to discriminate between very small energy differences, yet it is a sensitivity we routinely take for granted. It has been estimated that the human eye is capable of distinguishing between five and eight million different colors.

2.5 The Electromagnetic Spectrum

The large energy range of electromagnetic radiation from more than three million electron volts to less than one ten-thousandth of an electron volt is called the electromagnetic spectrum. The human eye is sensitive to just a very small portion of the spectrum corresponding roughly to the energy-range output of the sun that bathes the earth; sensitivity to other wavelengths would not be useful to the eye. However, other wavelength ranges do exist. This was demonstrated initially by Sir William Herschel (1738–1832), a German-born British astronomer, who dispersed achromatic light and then placed a thermometer covered with an opaque material in the region beyond the red region—and the temperature of the thermometer rose! Herschel had discovered the infrared (IR) region of the spectrum, a region with less energy per wave (photon) than the red region, where there seems to be more heat than light. He reported his findings in a series of articles in 1800 [5–8]. Herschel is most famous for his astronomical discoveries, particularly of the planet Uranus as well as two of their moons. He worked closely with his sister, Caroline Herschel (1750–1848), who was an accomplished astronomer in her own right.

A corresponding experiment on the violet end of the spectrum was performed by Johann Wilhelm Ritter (1776–1810), a German chemist and physicist. Ritter placed silver chloride crystals, that were known to darken on exposure to light, in

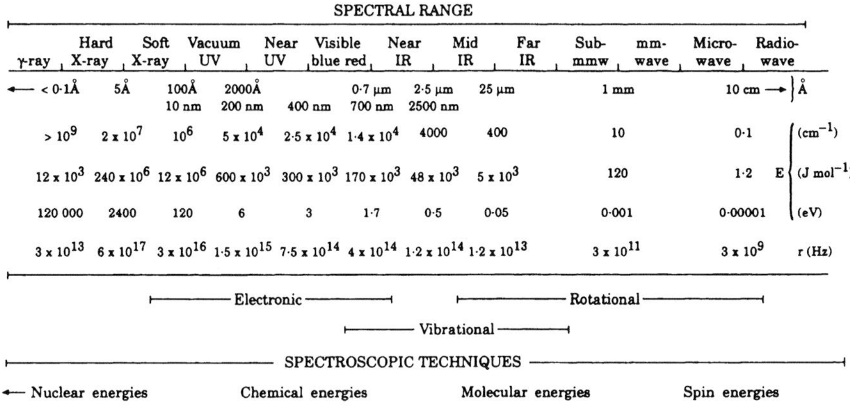

SPECTRAL RANGE

γ-ray	Hard X-ray	Soft X-ray	Vacuum UV	Near UV	Visible blue red	Near IR	Mid IR	Far IR	Sub-mmw	mm-wave	Micro-wave	Radio-wave	
← < 0·1Å	5Å	100Å / 10 nm	2000Å / 200 nm	400 nm	0·7 µm / 700 nm	2·5 µm / 2500 nm	25 µm		1 mm		10 cm →		$\big]$ Å
> 10^9	2×10^7	10^6	5×10^4	$2·5 \times 10^4$	$1·4 \times 10^4$	4000	400		10		0·1		(cm^{-1})
12×10^3	240×10^6	12×10^6	600×10^3	300×10^3	170×10^3	48×10^3	5×10^3		120		1·2 E		$(J\ mol^{-1})$
120 000	2400	120	6	3	1·7	0·5	0·05		0·001		0·00001		(eV)
3×10^{13}	6×10^{17}	3×10^{16}	$1·5 \times 10^{15}$	$7·5 \times 10^{14}$	4×10^{14}	$1·2 \times 10^{14}$	$1·2 \times 10^{13}$		3×10^{11}		3×10^9		r (Hz)

|—— Electronic ——| |—— Rotational ——|

|—— Vibrational ——|

SPECTROSCOPIC TECHNIQUES

←— Nuclear energies   Chemical energies   Molecular energies   Spin energies

Fig. 2.4 The electromagnetic spectrum (with energy decreasing from left to right). © 1998, M.V. Orna

the region beyond the violet and noticed that they darkened more quickly than when placed in any other region of the visible spectrum. In his own words [9, 10]:

> *On 22 February, I also encountered rays alongside violet in the colour spectrum of colours—outside it—by means of horn silver. They reduce even more strongly than violet light itself, and the field of these rays is very wide.*

Ritter recognized that there are rays beyond the violet end of the visible prismatic spectrum that blacken a chemical substance, but for him, the rays were not themselves an entity, but revealed the inherent polarity in light. Given the fact that he did not realize what he had discovered in terms of modern interpretation, and that his ontological system was one of symmetry, he expected a similar chemical effect on the red end of the spectrum in addition to the heating effect found by Herschel. In addition, his experiments were highly criticized in the decade following his publication, so it took a long time before he realized that there were higher energy waves beyond the violet in what we now call the ultraviolet (UV) region [11]. Although Ritter is best known among chemists for his discovery of the UV, physicists also hail the fact that he constructed the first dry cell battery in 1802 and a storage battery in 1803. His most important contribution to electrochemistry came in 1798: Ritter was the first to establish an explicit connection between galvanism and chemical reactivity. He was so excited by this discovery that he used his own body for many of the tests he conducted on electrical excitation of muscle and sensory organs. It has been speculated that these experiments exacted a very high price from him: he died at the young age of 33.

Other regions of the spectrum were gradually discovered and characterized so that now, in Fig. 2.4, we can see the entire electromagnetic spectrum laid out.

If we can't see the regions of the spectrum beyond the visible, why are they important? As we can see from the diagram, the other regions can be very useful. They include microwaves and radio waves, without which modern life would be inconceivable. They also include the high energy regions, which can be very dangerous. For example, white snow and bright sunlight can be lethal. Ultraviolet light bouncing off a white surface can be so intense that it burns the cornea of the eye. This can cause the epithelial cell loss seen in cases of photokeratitis or snow-blindness. Severe pain, blurred vision, photophobia, and temporary blindness result. While most people recover within 24 h, prolonged exposure to reflected UV light can lead to some permanent vision loss. In our time, the health risk is exacerbated by ozone depletion in the atmosphere, leading to increased high energy UV radiation, and therefore increased damage to the eyes, the immune system and the skin [12]. Ultraviolet radiation can be harmful to more than the human eye: it can discolor virtually any color of organic origin by destroying the constituent molecules. High energy radiation can also be useful as diagnostic tools when used with caution. What lies beyond the infrared can be useful as well, and we shall see how very useful when we examine some of the major analytical techniques that help us understand and conserve colored artifacts.

2.6 How We See Color

At this point, we have seen the experimental evidence and the theories that developed around the nature of light and the fact that so-called white, or achromatic, light is anything but! We must now ask: how is it that we see the colors inherent in white light without the use of a modifying prism, and more importantly, how is it that we can see any color at all? To answer these questions, we must look at the three components necessary for the perception of color: the source of light, the object being observed, and the observer.

2.6.1 The Light Source

Every source of illumination emits a range of energies which vary across the energy spectrum to yield what is known as a spectral energy distribution curve. Figure 2.5, below, is the complete spectral energy distribution curve of sunlight from 0 to 3,000 nm when the sun is at its zenith (directly overhead). Measurements at other angles generate a family of these curves. Figure 2.5 contains a wealth of information, and some that can be inferred. First of all, we note that the radiation maximum (shown as irradiance measured in $W\,m^{-2}\,nm^{-1}$) occurs at about 500 nm and falls off sharply on both sides of the maximum. The detection limits of the human eye exactly match the maximum in this curve, indicating an evolutionary link between the development of vision and the environment. Other animal species, e.g., the butterfly, have different sensitivity ranges. Indicated also on this curve are the absorption bands of atmospheric components, ozone (O_3), water (H_2O) and carbon dioxide (CO_2). The fact that ozone absorbs a great deal of the high energy radiation below 400 nm demonstrates its protective nature in the atmosphere and why the so-called

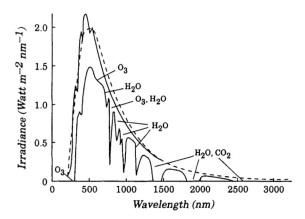

"ozone hole" over the Antarctic is an environmental catastrophe caused mainly by the presence of halocarbon refrigerants in the stratosphere. An extended consideration of this curve also gives us a clue as to why the sky is blue during the day and why many sunsets are brilliant red in color. Light is scattered in direct proportion to its energy, so blue–violet light is scattered by the particles in the atmosphere much more than the lower energy wavelengths of the solar spectrum, so when we look at the sky, we see the scattered blue light, and when we look at the sun, we see the sun's emitted light minus the blue, and hence blue's complementary color, yellow [13]. At the end of the day, when the sun is "setting," the sun's rays must travel through a greater air mass than at midday. Much of the sunlight is absorbed by the atmosphere, again in proportion to the energy of the waves, so the violet, blue, green and much of the yellow region of the solar spectrum is absorbed, leaving us to feast our eyes on a gorgeous red sunset!

As previously noted, a light source which emits energy with roughly constant radiant power over the limited response range of the human eye, 1.7–3.1 eV, or in terms of wavelength, 700–400 nm (1 nm, nm $= 10^{-9}$m), is perceived by the eye as "white."

Dispersion of this light by a dispersing instrument such as a prism or a grating yields the spectral colors ranging from red at around 1.7 eV to violet at around 3.1 eV. The light source, usually a luminous body like the sun, emits packets of energy called photons with a range of energies, and the intensity of the radiation may vary with the energy, yielding spectral energy distribution curves such as those shown in Fig. 2.6. The x-axis, as in Fig. 2.5, is wavelength measured in nanometers; the longer the wavelength, the lower the energy of a given photon. The color range is from violet at 400 nm through green at 500 nm to red at about 700 nm. Figure 2.6a shows the curve for typical sunlight restricted to the 400–700 nm range (just a portion of Fig. 2.5), Fig. 2.6b for a 100 W incandescent light bulb, and Fig. 2.6c for a 15 W standard cool fluorescent light bulb.

From these curves, we can see that the intensities of radiation for sunlight vary only a little over the entire wavelength range for visible light. The incandescent light bulb's radiation intensity increases dramatically toward the red end of the

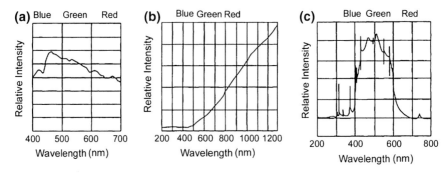

Fig. 2.6 Spectral energy distribution curves of **a** Sunlight, **b** Incandescent light, and **c** Fluorescent light. © 1998, M.V. Orna

spectrum (>700 nm), whereas its intensity is very weak in the violet–blue–green region. Incandescence is a phenomenon observed when objects are heated, and the hotter the object, the closer its radiant incandescent energy approaches white light. The radiation intensity of the fluorescent bulb exhibits a maximum at around 500 nm, but includes appreciable intensities of the entire wavelength range of the visible spectrum. A fluorescent lamp is actually a tube containing a gas at very low pressure that by application of high voltage, produces excited atoms that subsequently excite phosphors coating the inside of the tube; the phosphors convert the high energy UV light to a satisfactory range of visible light. So perhaps it should more properly be called a phosphorescence lamp.

A source which emits energy continuously over the limited response range of the human eye, from about 380 to 720 nm, and with appreciable intensities at all wavelengths, like sunlight, is perceived by the eye–brain complex as white and is therefore described as "white" or achromatic light. A source which emits energy with great intensities in the red region of the spectrum, and very little in the low wavelength region (blue–green region) of the spectrum is perceived as reddish–yellow, like an incandescent light bulb, commonly described as "warm" light. On the other hand, fluorescent lighting is very poor in red and is sometimes described as "cool" light for that reason; few people would appreciate fluorescent lighting in their living rooms. Yet for energy saving compact fluorescent lighting is now recommended, and will soon be mandated.

2.6.2 Interaction of Light with Matter

At this point, it is necessary to ask how light from the sources examined in the previous section might behave when it strikes an object, whether that "object" be the atmosphere, a liquid, solid or gaseous substance, or any other material substance one can think of. We can call any of these substances or objects the light "modifier" since light is indeed changed upon interaction. Some of the more common changes that light undergoes are: reflection, transmission, refraction, dispersion, scattering, absorption, diffraction, polarization, and interference. Each of these interactions can give rise to a perceived color change, and understanding the mechanism for each of them is important when studying the physics of color. Most important to the chemist are the phenomena of reflection, transmittance, refraction, and absorption, especially when dealing with colored objects.

2.6.3 The Object Observed

We all know that a light beam can be modified with respect to the direction in which it travels by being reflected from the surface of an object. In certain special cases, not only the appearance of an object, but also its color, can be affected by the manner in which it reflects light. In most instances, our experience of reflection involves our observation of reflection from a smooth surface such as a mirror or a pool of water or a polished metal surface. In these cases we always expect to see some sort of reflected image, and the integrity of the image depends upon the smoothness of the reflecting surface. The reason why is that light emitted by an object or a source as parallel rays will be reflected from an object at an angle equal to the angle of incidence, and the reflected rays all remain nearly parallel because each ray has a normal plane parallel to the normal planes of all the other rays. In the case of reflection from a rough surface, on the other hand, normal planes must be constructed perpendicular to the surface which a particular ray is striking, and these are seldom parallel. Since the reflected rays are no longer parallel either, very few rays from the same surface region will reach the observer's eye, so the surface will appear quite dull. This phenomenon explains the difference between a glossy finish and a matte finish. Furthermore, if the surface is very rough, it is possible that hardly any reflected light at all will reach the observer. Such a phenomenon is observed when platinum is precipitated as finely divided particles with such a rough surface that it is incapable of reflecting much light to the observer's eye and the material is called "platinum black," which may be a counter-intuitive term to those who only know platinum as a shiny, almost colorless, metal. Other important interactions of light with bulk matter are refraction, absorption and transmittance. For our purposes, refraction is a most important interaction. Some detailed refractive interactions are shown in Fig. 2.7.

In the seventeenth century, Christiaan Huygens had the insight that the various colors of light travel at different velocities in different media, and therefore, each color has a different angle of refraction. Although he formulated this principle in 1678, his treatise on it was not published until 1690. He proposed that each point in a wave of light can be thought of as an individual source of illumination that produces its own spherical wavelets, which all add together to form an advancing wavefront. This multiple wavelet concept is now known as the Huygens' principle [14]. Besides wave theory, Huygens is known for his eclectic interests in such things as the mechanics of clocks, probability theory, optics and astronomy. His

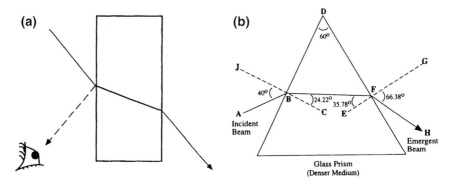

Fig. 2.7 Reflection and refraction at an acute angle to the reflecting surface for a transparent object with parallel sides (**a**) and detailed mode of interaction of light with a glass prism (non-parallel sides) (**b**). © 1998, M.V. Orna

rather heady circle of friends and colleagues included Blaise Pascal (1623–1662), René Descartes (1596–1650), and Giovanni Cassini (1625–1712).

Huygens' view is the only one that can help us understand refraction, the bending of light at the interface of two media with different densities. In both Fig. 2.7a and b, the light in air is shown entering and leaving a more dense medium, glass. In Fig. 2.7a, the glass plate has parallel sides. At the entrance surface, the light beam is bent toward the normal plane (an imaginary plane drawn at right angles to the surface) and at the exit surface, it is bent away from the normal plane, and the light beam resumes its original path, but slightly displaced. The reason why this is so is because the beam of light is a wave front and all the waves do not strike the surface at the same time when the angle of incidence is other than 90°, so the velocity of each wave does not change simultaneously, thus bringing about the bending we observe. When the object through which the light travels does not have parallel sides, as in the prism in Fig. 2.7b, then the entrance angle and the exit angle reinforce one another, bringing about the dispersion of each of the wavelengths that Newton observed and described. The shape of the prism is such that the light beam cannot resume its original path.

Two additional important interactions of light with matter are transmittance and absorption. In transmittance, as the word implies, the light passes through an object; the object must be transparent for this phenomenon to occur, but not necessarily colorless. We all know from our experience with stage lighting and colored filters that white light, when passed through a red filter, exits as red light. So, not only has transmittance occurred, but also the process of absorption, i.e., some of the light has been absorbed; only the red light was permitted to pass through the red filter; the rest of the light was held back, absorbed in the filter. So with a colored filter, we have observed transmittance and absorption at the same time. When light strikes an opaque object, this phenomenon of selective absorption can also take place, but what reaches our eye this time is reflected light. Let us see if we can describe this event more exactly. Figure 2.8a is the spectral energy distribution curve of ordinary sunlight. If this light strikes an opaque object that we

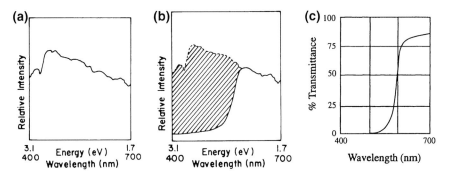

Fig. 2.8 Spectral energy distribution curve of ordinary sunlight (**a**), absorption spectrum of a red opaque object (**b**), transmittance spectrum of a transparent red object (**c**). © 1980, American Chemical Society, Ref. [15]

describe as red, then all the colors of the spectrum except red will be absorbed, and it will be the remaining red wavelengths that reach our eye.

If the light described by the curve in Fig. 2.8a were allowed to fall on an opaque object which absorbed some of the light, as shown in Fig. 2.8b, the light reflected to our eyes would no longer consist of significant intensities of all the wavelengths of visible light. The shaded area in the diagram is called the absorption band, and the unshaded area is the resulting reflectance curve of this "red" object. The light in the shaded area, which is largely green and blue light, has been absorbed to a great extent. Our eyes can then be stimulated only by the unabsorbed light at the red end of the spectrum, and so the object that yields this reflectance curve is perceived by the eyes as "red." The color characteristics of most colored objects can be described partially by reference to the shape, width, intensity and position of their respective absorption bands. The superimposition of the spectral energy distribution curve of Fig. 2.8a on the reflectance curve of Fig. 2.8b yields a composite curve called the "stimulus for color" curve, which stimulates the eye–brain mechanism to see color [8, 16]. Figure 2.8c is the transmittance spectrum of a hypothetical transparent red object—this is the type of spectrum that is usually measured for educational purposes by some simple spectrophotometers.

Color, however, is a very complex phenomenon. Objects can modify light not only by reflectance and selective absorption, as we have seen, but also by transmission, scattering, dispersion, interference, etc.—sometimes all at once. It is the combination of all these possible interactions which ultimately determine the appearance of an object.

2.6.4 The Eye–Brain Detector-Interpreter

After modification the light must strike a detector in order to be evaluated. The most important detector when discussing color is the human eye because perceived

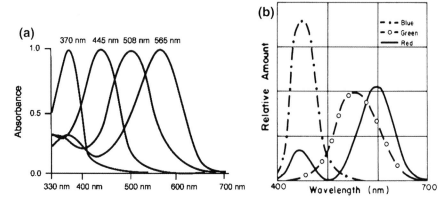

Fig. 2.9 **a** Sensitivity curves for the rods (370 nm curve) and cones (445 nm curve—blue; 508 nm curve—green; 565 nm curve—red, orange, yellow) in the human eye; **b** The 1931 CIE standard observer. © 1980, American Chemical Society, Ref. [15]

color is nothing more than the subjective personal evaluation of the light reflected or transmitted to the eye. A complete description of the color perception process must then involve the stimulus for color curve superimposed on the proper response curve for the human eye. Thomas Young (1773–1829) was the first to correctly recognize that color sensation is due to the presence of structures in the retina which respond to three colors. Young suggested that color blindness is due to the inability of one of these structures to respond to light. Hermann von Helmholtz (1821–1894) and James Clerk Maxwell (1831–1879) elaborated Young's work into a proper theory. We now know from anatomical studies that the retina of the eye contains cone-like structures that are sensitive to the major regions of the spectrum, roughly divided into red–green–blue, and rod-like structures that are not color-specific. Rods work in dim light and enable the mind to sense brightness. The cones work in normal intensity of light and allow the mind to sense colors. The eye works like a camera with two films, one for black and white, and one for color. A photochemical process selects the proper "film" for the correct lighting conditions [17]. Figure 2.9a shows the sensitivity curves intuited by Young, but in more detail.

The retinal rods are more sensitive in the ultraviolet, and there are specialized cones for the blue, green and red regions respectively. Since these sensitivities are slightly different for each human being, the 1931 Commission Internationale de l'Eclairage (CIE) defined the response curve for a "standard observer" in order to overcome this difficulty. This curve, which is illustrated in Fig. 2.9b is actually three curves, one for each response region of the spectrum, and it is based upon the Young-Helmholtz theory discussed above [18, 19].

For reasons that will be discussed in the following chapter, ultraviolet and visible spectra give rise to characteristic broad bands of radiation. When these broad bands correspond to each of several different regions of the visible spectrum, they are capable of inducing a mental color response interpreted as a single color.

Table 2.2 Colors of absorbed light and corresponding complementary colors

Wavelength (nm)	Energy (eV)	Color of absorbed light	Color seen
400–420	3.10–2.95	Violet	Green–yellow
420–450	2.95–2.76	Violet–blue	Yellow
445–490	2.76–2.53	Blue	Orange
490–510	2.53–2.43	Cyan	Red
510–530	2.43–2.34	Green	Magenta
530–545	2.34–2.28	Green–yellow	Violet
545–580	2.28–2.14	Yellow	Violet–blue
580–630	2.14–1.97	Orange	Blue
630–720	1.97–1.72	Red	Cyan

For example, if red, green and blue lights are mixed in the proper proportions, the mental color response of the human eye is "white." If blue light is subtracted from this mixture, and only the red–green combination remains, the human eye interprets this combination as "yellow." Together, blue and yellow "complete" the visible spectrum; thus they are termed complementary colors. When a chemical substance absorbs the wavelengths of blue light from a "white" light source, the remaining wavelengths will be reflected to the eye and interpreted as the color yellow [20]. Newton himself first recognized these relationships and organized the spectral colors into a color circle. When two colors directly opposite one another in the circle were mixed in equal proportions the result was white (considered to be the center of the circle). This view leads to an infinite number of complementary colors, and a number of variations on Newton's original color circle are in use today [2, 21]. Table 2.2 is a rough rendition of Newton's original color circle in tabular form.

2.6.5 Primary Colors

As we have seen in Fig. 2.9, the human eye has three types of receptor cones, a long wavelength red, a medium wavelength green, and a short wavelength blue. Because of the sensitivity of these cones over a broad wavelength range, most of the other colors perceived can be created by adding the wavelengths of these three colors. Thus, adding together red light, green light and blue light produces white light—each of the three regions of the visible spectrum taken together complete it, and each ranges over approximately one-third of the spectrum. These three colors, because of these characteristics, are called the additive primary colors. They are not mixtures of any other colors. Yellow, however, is a mixture of red and green lights; cyan is a mixture of blue and green lights; magenta is a mixture of red and blue lights. The three mixed colors, yellow, magenta, and cyan are called the subtractive primary colors and are mainly useful when dealing with opaque substances such as pigments that selectively absorb broad bands of visible light. Since

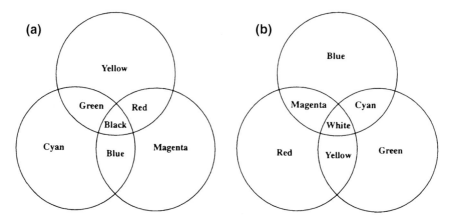

Fig. 2.10 Color circles of the subtractive primaries (**a**) and the additive primaries (**b**). © 1998, M.V. Orna

one diagram is worth a thousand words, let us take a look at Fig. 2.10 which are color circles that distinguish easily between the two types of primaries. The circles of Fig. 2.10a apply to the mixing of opaque colors such as paints. If yellow and cyan are mixed together, the yellow subtracts out the blue region of the spectrum, reflecting only red and green; the cyan subtracts out the red region of the spectrum, reflecting only blue and green. Since the only color that is reflected by both pigments is green, then the eye perceives the mixed color as green. Mixing all three subtractive primaries together produces black (or really in practice a muddy brown). The color directly across from each primary is its complement. Thus adding together yellow and blue pigments completes the subtractive process—all regions of the spectrum are absorbed [21].

The circles of Fig. 2.10b apply to the mixing of colored lights, such as one sees in stage settings—producing lights of different colors by using different filters. Mixing together red and blue lights produces magenta. Magenta is the complement to green. By adding green to magenta (actually red and blue) completes the spectrum and one sees a near facsimile of white light. So whenever primary colors are discussed, it must be made clear which ones: additive or subtractive?

One very practical application of these color circles is explaining why water appears blue when viewed in bulk. Due to hydrogen bonding, water absorbs very strongly in the infrared region and very weakly in the red–orange region of the spectrum. This slight red–orange absorption is enough to give water the bluish tint of the red–orange complement which is blue. Bulk ice in glaciers and icebergs also exhibits this blue color due to hydrogen bonding [22]. I was once standing on a glacier when someone asked our guide, "Why is the ice so blue?" The guide replied, "That is due to the ice reflecting the blue sky." I promptly made a mental note to verify anything else that the guide told us after that incident.

Figure 2.11 is a photograph of a freshly calved iceberg from the South Sawyer Glacier in Tracy Arm, Alaska. I leave it to the viewer to judge if the intense blue of

Fig. 2.11 Freshly calved iceberg. Photograph: M. V. Orna

this ice arises merely from reflecting the blue sky, which incidentally, was quite cloudy on the day the photograph was taken. An understanding of the physical basis of color production and the conditions needed to perceive color will be helpful as we move into the next phase of understanding color: what types of substances are colored, why are they colored, and how can we control color production?

References

1. Zollinger H (1999) Color: a multidisciplinary approach. Wiley-VCH, New York, p 11
2. Zollinger H (1999) Color: a multidisciplinary approach. Wiley-VCH, New York, pp 63–78
3. Newton I (1704) Opticks: or a treatise of the reflexions, refractions, inflexions and colours of light. Also two treatises of the species and magnitude of curvilinear figures. Royal Society, London. Reprinted in: Hutchins RM (ed) (1952) Great books of the western world, vol 34. Encyclopedia Britannica, New York, pp 386–412
4. Aristotle (1908) On sense and the sensible. The Clarendon Press, Oxford (Sect. 2 and 3)
5. Herschel W (1800) Investigation of the powers of the prismatic colours to heat and illuminate objects. Phil Trans Roy Soc Lond 90:255–283
6. Herschel W (1800) Experiments on the refrangibility of the invisible rays of the sun. Phil Trans Roy Soc Lond 90:284–292
7. Herschel W (1800) Experiments on the solar, and on the terrestrial rays that occasion heat, Part I. Phil Trans Roy Soc Lond 90:293–326
8. Herschel W (1800) Experiments on the solar, and on the terrestrial rays that occasion heat, Part II. Phil Trans Roy Soc Lond 90:437–538
9. Ritter JW (1801) Beobachtungen zu erwähnen die Entstehung von Thermoströmen. Gilbert's Annalen 7:527
10. Ritter JW (1802) Versuche über das Sonnenlicht. Gilbert's Annalen 12:409–415
11. Frercks J, Weber H, Wiesenfeldt G (2009) Reception and discovery: the nature of Johann Wilhelm Ritter's invisible rays. Stud Hist Philos Sci 40:143–156
12. Longstreth J et al (1998) Health risks. J Photochem Photobiol B 46(1–3):20–39
13. Minnaert M (1954) The nature of light and color in the open air. Dover, New York, p 238
14. Huygens C (1690) Traité de la lumiere. Pieter van der Aa, Leyden

15. Orna MV (1980) Chemistry and artists' colors: part I. Light and color. J Chem Educ 57:256–258
16. Billmeyer FW Jr, Saltzman M (1981) Principles of color technology, 2nd edn. Wiley, New York, pp 2–12
17. Orna MV (1978) The chemical origins of color. J Chem Educ 55:478–484
18. Wald G (1964) The receptors of human color vision. Science 145:1007–1016
19. Wasserman GS (1979) The physiology of color vision. Color Res Appl 4:57
20. Davis JC Jr (1975) Light and the electromagnetic spectrum. Chemistry 48(5):19–22
21. Küppers H (1979) Let's say goodbye to the color circle. Color Res Appl 4:19
22. Nassau K (1983) The physics and chemistry of color. The fifteen causes of color. Wiley-Interscience, New York, p 72

Chapter 3
The Chemical Causes of Color

Thousands of tourists from all over the world flock to two island sites at opposite ends of the earth from one another for no other reason than to gaze in awe and wonder at the color phenomenon each one is famous for. Hawaii's Green Sand Beach on the southernmost point of the Big Island boasts an entire beach covered with tiny olivine crystals that sparkle like emeralds in the tropical sun. Olivine is a common constituent of lava flows, but only in this one place has it been found of gem quality—the crystals in this case seem to have floated on a former lava lake [1]. Around the globe in the North Atlantic, a unique body of water known as the Blue Lagoon on Iceland's Reykjanes peninsula plays host to about 400,000 visitors a year who come to take a dip in this unique body of water—also an indirect product of an ancient lava flow flooded by the transforming effluent from a nearby geothermal power plant [2]. What is it about some materials that give rise to these and so many other wonderful colors? Why are other substances colorless, and can we modify them to produce color? The answers to these questions are not simple, and they involve awareness of several important insights regarding the nature of matter, particularly atoms and molecules.

3.1 Development of Atomic Theory

While nineteenth century physicists were busy exploring the nature of light and energy, other scientists who would later be called chemists were concerned with the nature of matter—what it was composed of and how it could be transformed. Speculative ideas about the particle nature of matter came originally from the Greeks, but were given a firmer basis by English schoolmaster John Dalton (1766–1844), who asked in 1803 [3], "Why does not water admit its bulk of every kind of gas alike? This question I have duly considered, and though I am not able to satisfy myself completely I am nearly persuaded that the circumstance depends on the weight and number of the ultimate particles of the several gases." From this breakthrough statement Dalton went on to hypothesize, always based on

M. V. Orna, *The Chemical History of Color*, SpringerBriefs in History of Chemistry, DOI: 10.1007/978-3-642-32642-4_3, © The Author(s) 2013

Table 3.1 Timeline of discoveries leading to modern atomic structural theory

Discovery	Date	Discoverer
Atomic theory	1803	John Dalton (1766–1844)
Periodic properties of the elements as an apparent function of their atomic weights	1869	Dmitri Mendeleev (1834–1907)
Cathode rays and their negative nature (electron)	1879	Sir William Crookes (1832–1919)
Balmer series	1885	Johann Balmer (1825–1898)
Canal rays and their positive nature (proton)	1886	Eugene Goldstein (1850–1930)
X-rays	1895	Wilhelm Roentgen (1845–1923)
Phenomenon of radioactivity	1896	Henri Becquerel (1852–1908)
Electron (Crookes 'cathode rays') charge/mass ratio	1897	J. J. Thomson (1856–1940)
Alpha, beta, and gamma "rays"	1898	Ernest Rutherford (1871–1937)
Planck's Law (E = hν)	1900	Max Planck (1858–1947)
Photoelectric effect	1905	Albert Einstein (1879–1955)
Atomic nucleus	1910	Ernest Rutherford (1871–1937)
Isotopes	1913	Frederick Soddy (1877–1956)
Bohr model of atomic structure	1913	Niels Bohr (1885–1962)
Atomic number	1914	H. G. J. Moseley (1887–1915)
Neutron	1932	James Chadwick (1891–1974)

experiment, which distinguished him from his Greek predecessors, that matter was composed of particles differentiated from one another by their weight, and that they combined with one another in simple proportions. It would be more than a century before the structure of these particles, which we now call atoms, was even partially understood. Some more or less detailed timelines regarding the development of atomic structure can be found on the internet [4]. Table 3.1 is an abbreviated version that will help us to get to where we want to go quickly. The tabular form helps us realize that a great flurry of activity took place in the last two decades of the nineteenth century and the first of the twentieth century that has had profound effects on our theoretical framework in physics and chemistry ever since.

Once the particulate nature of matter began to evolve, it was a matter of time (over half a century) before a coherent arrangement of the different types of atoms based on their atomic weight began to take shape (Mendeleev [5]). A decade later, one of the fundamental subatomic particles was discovered, the electron (Crookes), but this discovery was not understood for what it was until Thomson [6] measured its mass-to-charge ratio almost 20 years later. Meanwhile, Eugene Goldstein [7] discovered canal rays, which were positively charged, but again, the significance of this discovery did not "register" until Rutherford's nuclear model [8] of the atom appeared. Meanwhile, the combined discoveries of Planck's Law [9] and the photoelectric effect [10] (Einstein) allowed physicists for the first time to step away from the model of the classical oscillator and embrace the idea that energy, as well as matter, was discrete, i.e., that energy was taken up and given off by matter in "quanta," and not continuously, leading to the modern quantum theory. Then in 1914, Moseley's work demonstrated that it was the atomic number [11], eventually associated with the number of protons present in an atom, that differentiated atoms from one another rather than the atomic weight. In 1913, Frederick Soddy realized the existence of isotopes [12]; the final piece of the atomic structure puzzle fell into place in 1932 with James Chadwick's discovery of the neutron [13]—so now isotopes could make sense! The finely woven threads of the tapestry describing atomic structure were eventually tied together by an evolving understanding of the nature of X-rays, radioactivity, and the "rays" emitted by radioactive elements. Bohr's model of atomic structure [14], that postulated discrete energy levels within the atom, was an "aha!" moment derived from an explanation of Johann Balmer's 1885 empirical formula [15] for calculating the visible wavelengths in the hydrogen spectrum.

3.2 The Chemical Bond

From this series of discoveries the modern picture of the structure of the atom emerged, and during the first half of the twentieth century, chemists like Walther Kossel (1888–1956), Irving Langmuir (1881–1957) [16], G. N. Lewis (1875–1946) and Linus Pauling (1901–1994) developed our present understanding of the nature of the chemical bond.

Regarding the chemical bond, we must turn the clock back to 1850 in order to give credit where credit is due. In that year, Edward Frankland (1825–1899), a British chemist best known for his work on water quality and analysis, developed a theory of valence following his research into the properties of $Zn(CH_3)_2$, zinc methyl in Frankland's terminology, a new reactive organometallic compound, and a series of alkyl-conjugated metals that had different combining powers than the metals alone. Colin Russell remarks [17]:

> One of the most fundamental doctrines of chemistry has been that of valency (valence), second only perhaps to the atomic theory upon which it is founded. Its early history has been well-charted and few would now deny that Frankland was one of its major architects....there can be little doubt about the origins of valency theory. They lay in Frankland's own work in organometallic chemistry.

In Frankland's own words from the paper he subsequently published in 1852 [18],

Fig. 3.1 Lewis diagram
taken from his original paper,
Ref. [20]

I had not proceeded far in the investigation of these compounds...before the facts brought to light began to impress upon me the existence of a fixity in the maximum combining value or capacity of saturation in the metallic elements which had not before been suspected.

This theory, which we now call valency, has dominated the subsequent development of chemical doctrine and forms the groundwork upon which the fabric of modern structural chemistry reposes. Less than a decade later, the idea of valence bore fruit in the structural work of A. W. von Hofmann (1818–1892), Archibald Scott Couper (1831–1892), and Friedrich August Kekulé (1829–1896), the need for which arose out of the burgeoning dye industry following the discovery of William Henry Perkin (1838–1907). This development is one of the subjects discussed in, Sects. 5.2 and 5.4.

Almost simultaneously, in 1916, Walther Kossel and G. N. Lewis published their ideas on atomic combination. Kossel was mainly concerned with ionic bonding [19]:

Each valence position as seen chemically has a 'valence electron' arranged on the atom, but the specific negative valence position, that on the halogen in our case, is perhaps to be considered as a 'free place' with respect to the element following in the periodic system, which the atom strives to fill up.

Lewis, on the other hand, eschewed anthropocentric terminology and concerned himself with the behavior of the valence electrons being held in common by two atoms [20] (Fig. 3.1).

We may give a complete formula for each compound by using the symbol of the kernel instead of the ordinary atomic symbol, and by adjoining to each symbol a number of dots corresponding to the number of electrons in the atomic shell. It is evident that the type of union which we have so far pictured, although it involves two electrons held in common by two atoms, nevertheless corresponds to the single bond as it is commonly used in graphical formulae.

Irving Langmuir enlarged upon and summarized these ideas into a formulation that we now know as the "octet rule" [21]:

The properties of the atoms are determined primarily by the number and arrangement of electrons in the outside shell and by the ease with which the atom is able to revert to more stable forms by giving up or taking up electrons...The most stable arrangement of electrons is that of the pair in the helium atom...The next most stable arrangement of electrons is the octet, that is, a group of 8 electrons like that in the second shell of the neon atom. Any atom with atomic number less than 20, and which has more than 3 electrons in its outside layer tends to take up enough electrons to complete its octet.

Almost immediately, these heady ideas were taken up by Linus Pauling while he was still an undergraduate student. By 1920, when he was a senior, he was offering a seminar on the electronic nature of the chemical bond, basing the course on the ideas cited above. Beginning in 1922, Pauling attended the California

Institute of Technology (Caltech). At the time Caltech was one of the first institutions to use the new methods of X-ray crystallography to analyze structures, and while still a graduate student, Pauling published seven papers on crystal structures. By 1928, after studying in Europe for a year, he was doing frontier research on chemical bonding. Beginning in 1931 he published a seven-part series entitled "The Nature of the Chemical Bond" [22–29]. New rules for determining bond lengths and bond angles, magnetic moments and other molecular properties were described using the concept of resonance. Pauling's theory held that molecules can be represented by a linear combination of wave functions. He thus transformed the field of chemistry by applying quantum theory and quantum mechanics to chemical structure and bonding. In 1954, he received the Nobel Prize in chemistry for this work.

3.3 Electronic Transitions: The Hydrogen Spectrum

The concepts of molecular structure developed from the theory of atomic structure provide us with a picture of vibrating nuclei linked by electrons located in permitted orbitals of different energies. The consolidation of the quantum theory by the successful interpretation of the energy distribution of blackbody radiation, the line spectra of atoms, and the band spectra of molecules, leads us to believe in the existence of permitted energy levels. Absorption of energy by these atoms, molecules and crystal systems serve to move a body from place to place (translational energy), to cause electrons in permitted energy levels of the system to enter higher energy levels (electronic energy), to allow the atoms in a system to change their distances with respect to one another (vibrational energy), and to allow gaseous molecular systems to rotate (rotational energy). The energies of visible light are too great to be absorbed as rotational or vibrational energy, but in many instances, they are sufficient to promote electronic systems from their ground state to excited states. An example may be drawn from the simplest atomic system, hydrogen, as shown in Fig. 3.2.

From the diagram, we can see that the energy necessary to promote hydrogen's single electron from the ground state (the lowest possible energy state, $n = 1$, where n is the lowest permitted energy level for hydrogen) to any of the other levels that have higher values of n, that is, to higher permitted energy levels, is at the very least a little over 10 eV. This is a value much greater than any light wave of visible light. This so-called Lyman Series (named after Theodore Lyman (1874–1954), a Harvard physicist who first discovered it in 1906) obviously represents energy transitions in the ultraviolet region of the electromagnetic spectrum. However, the transitions involving the relaxation of an electron from higher states to $n = 2$ fall within the visible region of the Balmer Series. On the basis of this single observation, it is now possible to answer our fundamental chemical questions. In the first instance, the process which allows some species to absorb visible radiation is an electronic process whereby radiation is absorbed to promote the

Fig. 3.2 Electronic energy
levels of hydrogen. © 1980,
American Chemical Society,
Ref. [30]

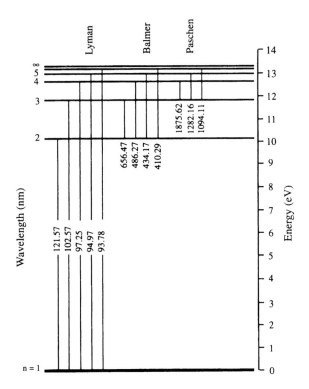

species into excited states, and radiation is emitted when relaxation from higher to lower energy states takes place. In the second instance, not all species can produce color when they undergo this process because their permitted energy level differences, which are determined by the structure of each species, lie outside the visible region. Thirdly, different materials exhibit different colors because the energy level spacings in their molecules, atoms and crystals are different. However, all molecules, atoms and crystals, whether colored or colorless, exhibit absorption of electromagnetic radiation. Since there is no difference in principle between electronic transitions resulting from absorption from the visible or ultraviolet regions, one can say that color is not connected with any one feature of molecular structure [30].

When an electronic transition takes place such as that illustrated in Fig. 3.2 for the lowest energy transition in the Balmer Series, 656.47 nm, a single color will be absorbed or emitted in the process. When viewed through a narrow aperture, such as a slit commonly used in rudimentary spectrometers, a narrow line of that color is seen as an image of the slit. The spectrum consists of a single monochromatic line. However, most materials, when they absorb or transmit visible radiation, exhibit a broad spectrum over a range of wavelengths because in a molecule, there are vibrational and rotational energy levels superimposed on the electronic energy levels, so a number of wavelengths on either side of the principal absorption band are also absorbed. These broad bands are characteristic of visible and ultraviolet spectra of molecules.

The remainder of this chapter will be devoted to examining the causes of color in organic and inorganic compounds. Due to the limitations of the scope and size of this book, only transitions between electronic energy levels in ions or molecules (in transition metal compounds, organic compounds, and charge transfer complexes) and transitions involving energy bands (in semiconductors) will be considered, although there is a much more comprehensive work available that also deals with vibrations and simple excitations, and geometrical and physical optics [31].

3.4 Color in Organic Compounds

Development of theories regarding the sharing of electrons led to the concept of the covalent bond, a particularly appropriate one for dealing with the carbon–carbon bonds found in almost all organic compounds. Linus Pauling's quantum mechanical calculations based on the idea of resonance led to an approach that came to be known as the valence bond (VB) approach, whereas another approach based upon linear combination of atomic orbitals to form molecular orbitals led to the molecular orbital (MO) approach. The latter model is very helpful in understanding the reasons why certain types of organic molecules are colored, and can even predict the intensities of absorption bands for simple molecules.

In 1876, Otto Nikolaus Witt (1853–1915) offered a qualitative relationship between the color and constitution of organic molecules. He suggested empirically that a colored molecule must contain an unsaturated group called a chromophore in order to exhibit color [32]. Some common chromophores are –N=N–, –C=C–, –C=O, and the phenyl group, although their presence does not necessarily give rise to color. He also proposed that the presence of other groups, called auxochromes, such as –OH, –NH$_2$, and –NHR, served to strengthen and deepen the color of a molecule. A molecule is called a chromogen when it contains a chromophore, but not an auxochrome. However, addition of auxochromes or accumulation of chromophores can lead to color development in a chromogen. The basis for these observations could not have been understood until the development of the quantum mechanical models named above.

3.4.1 The MO Model

Molecular orbitals can be thought to be generated by the overlap of atomic orbitals, which are wave functions that may reinforce the composite wave function (in-phase overlap) or may destructively interfere with it (out-of-phase overlap). In-phase overlap of *s* atomic orbitals or of end-on *p* orbitals is particularly effective

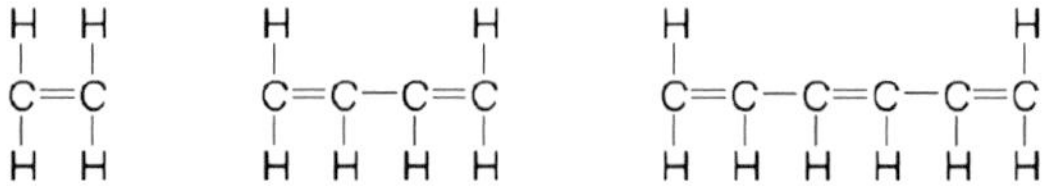

Fig. 3.3 Effect of extent of π-conjugation on the energy of excitation

and lead to sigma (σ) bonding MOs where the wave function has a larger magnitude in the internuclear region and are strongly bonding (of low energy). Conversely, out-of-phase overlap leads to high energy sigma ($\sigma*$) antibonding MOs with little probability of electron density in the internuclear region. Sideways overlap of p atomic orbitals also leads to bond formation but of the π type, which may be bonding (π) or antibonding (π*) as well. These bonds are more delocalized than sigma bonds and less tightly held by the molecular framework. Within the framework, there may also be nonbonding (n) electrons present. Energy absorption for this type of molecule can be via transitions of high energy ($\sigma \rightarrow \sigma*$), intermediate energy ($\pi \rightarrow \pi*$), or low energy ($n \rightarrow \pi*$). Generally speaking, $\sigma \rightarrow \sigma*$ transitions require high frequency radiation in the far UV and they do not come into play when constructing or predicting the types of molecules that will produce color. $n \rightarrow \pi*$ transitions require far less energy for molecular excitation and generally occur in the visible region. $\pi \rightarrow \pi*$ transition energies, at least for small molecules with limited conjugation, take on values intermediate to the other two types and occur in the near UV and visible regions. Extended π-conjugation leads to a bathochromic shift in the $\pi \rightarrow \pi*$ absorption maxima. For example, let us consider a series of molecules that contain the chromophore –C=C– by adding one –C=C– at a time to make ethene, 1,3-butadiene, and 1,3,5-hexatriene (Fig. 3.3):

We observe that as the –C=C– chromophore accumulates, the required transition energy decreases such that we would expect that a transition in the visible region would eventually become possible. Nature has actually given us such a molecule, containing 11 such chromophores: β-carotene, known to all for its orange color due to absorption in the blue region of the visible spectrum: its absorption maximum is 470 nm (about 2.5 eV) (Fig. 3.4).

An example of the effect of the presence of heteroatoms, especially those containing lone pairs of electrons, can be seen in comparing stilbene, which is colorless, with azobenzene, which is orange. The presence of the non-bonded electrons on the nitrogens in the azo compound allow for $n \rightarrow \pi*$ transitions, which occur for the most part in the visible region of the spectrum (Fig. 3.5).

3.4.2 The VB Model

This model is very familiar to chemists in its semi-intuitive qualitative extension known as resonance theory, which was first proposed by Linus Pauling in his work

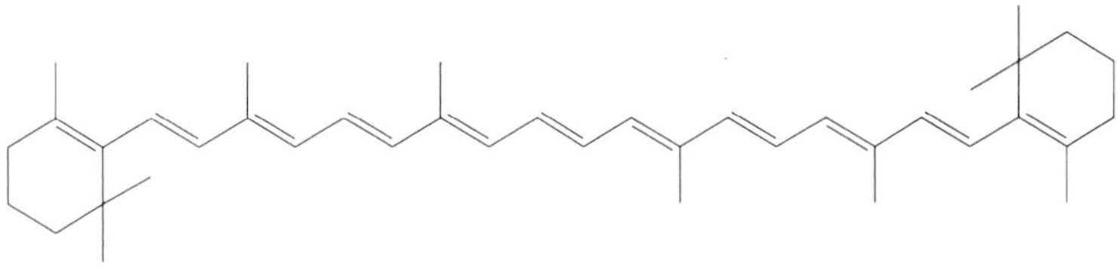

Fig. 3.4 Molecular structure of β-carotene

Fig. 3.5 Comparison of the
structures of stilbene and
azobenzene

Stilbene   Azobenzene

on the nature of the chemical bond [22–29, Parts V, VI and VII]. The shift to longer wavelengths (lower energies) that we observed with accumulation of chromophores is treated in this model in terms of an increasing number of contributing structures with similar or like relative stabilities. VB theory can predict the expected location of absorption bands.

3.4.3 The Free Electron Model

For a long-chain conjugated hydrocarbon with n carbon atoms in the chain, and a like number of delocalized π-electrons, we can assume that the π-electrons can freely range in a well of constant potential energy whose boundary is slightly longer than the length of the carbon chain itself. These assumptions simplify the system so that the solution for the Schrödinger equation for a particle in a one-dimensional box applies, allowing for an estimate of the theoretical absorption maximum that has been found to be very close to the observed values for several series of dyes [33, 35].

3.4.4 Modification of Witt's Terms and Classification of Colored Organic Compounds

Although very popular even today, Witt's color terms, chromophore, chromogen and auxochrome, have no theoretical definition as such. Some broader definitions of these terms that will help bridge the gap between empirical observation and semi-quantitative understanding are the following:

• Chromophore: any unsaturated grouping that is colorless (as in the examples given by Witt)

- Chromogen: an unsaturated system that is colored, or can be rendered colored, by attaching simple substituents
- Auxochrome: abandoned in favor of a less ambiguous term, "electron donor group"

Using these suggested terms, any atom that possesses lone pair electrons in conjugation with a π-electron system can be regarded as an auxochrome.

Using these revised definitions, one can conveniently classify the large diversity of colored organic molecules into four broad classes:

- $n \rightarrow \pi^*$ chromogens
- Donor–acceptor chromogens
- Acyclic and cyclic polyene chromogens
- Cyanine-type chromogens

The vast majority of colored organic compounds are based on donor–acceptor chromogens, and with the exception of the polycyclic quinones and the phthalocyanines, all the commercially important synthetic dyes are of this type [34, 35].

3.5 Colored Inorganic Compounds

It is often convenient, as it is now, to divide the fundamental ways in which color is created in chemical substances into two classes corresponding to the traditional division into organic and inorganic compounds. However, it will become apparent later in the discussion that this division is not a strict one since some compounds in both classifications exhibit color by the same type of mechanism. In all cases, electronic transitions between energy level differences in the visible region of the spectrum must take place.

3.5.1 Electronic Transitions in Coordination Compounds

Toward the end of the nineteenth century, compounds that defied the usual rules of valence occupied the thoughts of many inorganic chemists. In 1893, Alfred Werner (1866–1919) published his famous paper, "Contribution to the Constitution of Inorganic Compounds," [36] in which he set forth his ideas of coordination sphere and molecular geometry as applied to inorganic compounds. He postulated that coordination compounds consisted of a central metal cation around which were arranged either neutral molecules or anions, called ligands, and that each metal ion had a characteristic number of coordination sites which he termed the "coordination number." He also proposed that the bonds to the ligands were fixed in space and could therefore be treated by the application of structural principles. Although Werner's theory accounted for the structures of such compounds as

Fig. 3.6 Some Werner complexes with the labels written in Werner's own hand. University of Zurich Collection. Photograph courtesy of Roger Rea

$[Co(NH_3)_5(H_2O)]Cl_3$, it had nothing to say about the nature of the bonding within the coordination sphere (Fig. 3.6).

Linus Pauling made the first successful application of VB theory to coordination compounds, but the theory failed to explain why the elements in the d-block of the fourth row of the periodic table exhibit color in most of their compounds. These colors result from electronic transitions, but the inability of VB theory to account for these transitions, and for the resulting colors, was probably a key factor in the ascendancy of other views on bonding.

In the late 1920s and early 1930s, Hans Bethe (1906–2005) and J. H. Van Vleck (1899–1980) developed the crystal field theory of coordination compound bonding [37]. This theory considers the effect of the ligand on the electrons in the d-orbitals of the central metal ion. Common spatial arrangements of the ligands are octahedral, tetrahedral, and square planar, each of which distorts the spherical field of the fivefold degenerate d-orbitals such that the degeneracy is lifted, splitting the d-orbital energies into two or more energy levels. If the energy level difference, Δ, falls between 2 or 3 eV, then the metal ion-ligand complex should be capable of exhibiting color if there is an electronic population of at least one in the lower level, and at least one vacancy in the upper level. For example, $Ti(H_2O)_6^{3+}$ absorbs at 490 nm (in the green-yellow region); the complementary color, red-violet, is the color seen. Scandium, with no d-electrons, exhibits no color in any of its compounds; zinc, with its full complement of ten d-electrons, likewise has only colorless compounds. Since the d-block elements of the fifth and sixth rows of the periodic table have crystal field splitting energies from 25 to 50 % greater that the titanium to copper series, they absorb mostly in the ultraviolet region and are, for the most part, colorless. Many minerals and gemstones exhibit color due to crystal field splitting.

Fig. 3.7 Schematic of four-coordinate and six-coordinate metal ions complexed with a bidentate ligand. M represents the metal ion in these diagrams, and $\binom{x}{x}$ represents the ligand, where the line between the X's represents the entire middle portion of the ligand molecule

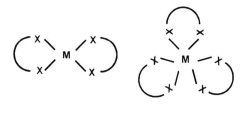

3.5.2 Color in Chelate Complexes

Whenever a ligand such as water coordinates with a given transition metal ion, it does so as an entity that normally has a single coordination site because it possesses only one donor group of electrons with the proper orientation. Examples of other such donors are the halogens, ammonia and the cyanide ion. Some organic ligands, however, possess more than one donor group and are capable of complexing with a metal ion in several configurations such as the following (Fig. 3.7):

Ligands, when viewed schematically, look something like a set of teeth, capable of biting into a metal ion. Thus, a ligand possessing two such coordination sites is said to be bidentate. Ligands having three, four, and six coordination sites are tri-, tetra and hexa- dentate respectively. When they complex with a metal ion, they clamp onto the ion with a tooth-like grip (hence the attribution of imaginary dentures to ligands), but also with a claw-like grip in the imaginative language of other chemists. Hence, though a mixed metaphor, these complexes in general are called "chelate" complexes from the Greek word for claw, *chela*. For example, when a bidentate chelating agent, 2,4-pentanedione, chelates with Cu(II), it forms a green square-planar complex (Fig. 3.8):

Many chelate complexes are highly colored, generally producing color in the same way as the ordinary transition metal complexes discussed in the previous section, as well as by pi–pi* and metal to ligand or ligand to metal charge transfer (discussed below). Some of the intensely colored compounds essential for life processes belong to this group of complexes. For example, chlorophyll is a chelate complex of magnesium, and heme, the action part of the hemoglobin molecule, is a chelate of iron. The cytochromes are other special chelates indispensable for life.

Of special interest are the chelate complexes that form when mordanting takes place in the dyeing process. One more metaphor associated with chelates: the Latin word *mordere* means "to bite," and a mordant is a substance that allows the dye molecule to bite into it. In so doing, the dye and the mordant form a new chemical substance as we see above in the reaction of the copper(II) ion with two molecules of 2,4-pentanedione. Using a mordant with a dye has some important effects:

- The dye is chemically bound to the mordant, which is also attached to the textile fibers, and thus the dye adheres more tightly to the textile

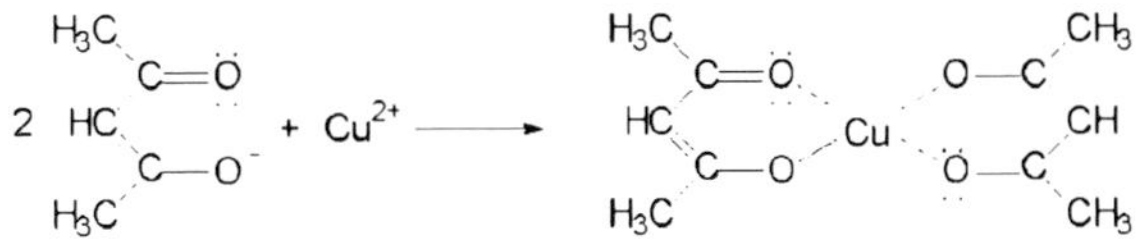

Fig. 3.8 Chelate formation between acetylacetone, a bidentate ligand, and copper ion

- The dye forms different chemical compounds with different mordants, so that a single dye can yield multiple colors when used with different mordants
- When the textile is changed, a different color will result even when using the same mordant and the same dye
- If the mordant and dye methods are harsh, then applying the mordant either before or after dyeing can limit potential damage to the textile

Some examples of mordants are salts of aluminum, copper, iron and chromium; in each case it is the metal ion that is the mordanting agent. A comprehensive table showing the effect of mordanting on plant dyestuffs from ancient Israel is in reference [38]. Numerous other chelates are known, and their usefulness spans the various branches of theoretical and applied chemistry and allied fields.

3.5.3 Charge Transfer Transitions

A host of inorganic colored compounds still has been left unaccounted for. For example, some mercury, bismuth and lead compounds exhibit brilliant colors, but they are not transition element compounds. Chromium as chromate is bright yellow and manganese as permanganate is deep purple, yet in neither of these cases do the metals have available d-electrons to be promoted to excited states. Since an absolute requirement of color production in chemical species is the presence of the proper and available energy levels, color in such compounds can be accounted for, at least in part, by recourse to the charge-transfer phenomenon.

In 1923, the Austrian physicist, Karl Przibram (1878–1973) first recognized that ordinary table salt in the gas phase will exhibit photon absorption at around 234 nm [39]. As a gas, table salt consists of a single sodium ion bound to a single chlorine ion by an ionic bond. The resulting ionic structure in the ground state may be represented by $Na^+ - Cl^-$ and the observed ultraviolet absorption band at 234 nm may be attributed to a transfer of an electron from Cl^- to Na^+ to form a covalently bonded NaCl molecule represented by $Na - Cl$. Many other salts related to sodium chloride exhibit charge transfer spectra in the ultraviolet or far ultraviolet regions and are examples of intramolecular transitions in inorganic compounds.

However, NaCl is colorless, as are many inorganic compounds exhibiting charge transfer spectra. On the other hand, those that absorb in the visible region yield highly intense and very dramatic colors. For example, when Fe(III) is placed in aqueous solution with excess chloride ion, it has the possibility of associating with the chloride ion to form the species $FeCl^{2+}$, $FeCl_2^+$, and $FeCl_3$. All of these species absorb intensely in the ultraviolet and the bands extend into the blue region of the spectrum, thus accounting for the yellow to orange colors of $FeCl_3$ solutions. These color-producing absorption bands have been attributed to the charge transfer phenomenon. Another iron complex that is color-producing is the iron-thiocyanate complex. Thiocyanate, CNS^-, tends to transfer its charge to Fe^{3+} to yield the complex $Fe^{2+}-CNS$, which absorbs intensely around 500 nm and exhibits a deep red color [40].

Considering any molecule or ion bound to a metal as a ligand, one can generalize the charge transfer phenomenon by stating that it occurs when an electron is transferred from an orbital lying principally on the ligand to an orbital lying principally on the metal, or vice versa. The former case is known as a ligand to metal or a L → M transition, and the reverse is a M → L process. The process may occur among the transition metal complexes as well as among the other elements; but only absorption occurring in the visible region will result in production of color.

Most charge transfer processes require higher energies than crystal field transitions and therefore usually lie in the ultraviolet or far ultraviolet regions of the spectrum. However, if the metal is easily oxidized and its ligand is easily reduced, then the M → L charge transfer transition may occur in the visible region. The reverse is also true. Charge transfer spectra are usually very intense and may mask any relatively weak crystal field interactions also present. Some well-known and intensely colored charge transfer compounds are the deep purple permanganate ion, MnO_4^-, and the yellow chromate ion, CrO_4^{2-}. The nature of the metal ion bound to these anions also is of consequence. For example, most chromates such as those of potassium and lead, are yellow, but silver chromate is red. Many organic compounds also exhibit charge transfer.

3.5.4 Semiconducting Materials

Materials like the cadmium compounds of sulfur, selenium, and tellurium, which are, respectively, yellow, red, and black, exhibit color by a mechanism we have not examined previously. Since cadmium has a d^{10} electronic configuration, crystal field splitting cannot be the origin of the color, but the color can be explained with recourse to a look at semiconductor properties.

There is a significant difference between color production by an ordinary charge transfer transition and color production in a semiconductor. In the latter instance, the electron to be promoted to a higher energy state does not occupy a discrete energy level in the ground state, but is a member of a delocalized population of electrons occupying a band of extremely closely spaced, and therefore indistinguishable, energy levels called the valence band. Absorption of a photon can promote it across the band-gap, the magnitude of which is determined by the nature of the semiconductor, to an excited state consisting of the electron population in the conduction band. Since the conduction band consists of numerous

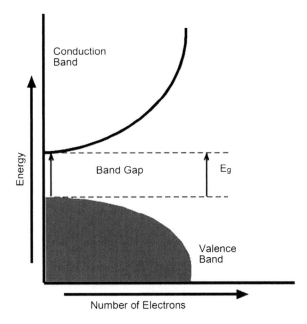

Fig. 3.9 Semiconductor behavior: energy transition across a band-gap. © 2011. M. V. Orna

closely-spaced energy levels, the exciting photons need only a minimum energy to overcome the band-gap, E_g, and photons of all energies above this minimum will be absorbed.

From Fig. 3.9 it can be seen that the minimum energy necessary to promote a semiconductor to an excited state is E_g. Photons possessing energies between 1.7 and 3.1 eV are those that lie in the visible range. Substances with E_g values greater than 3.1 eV cannot absorb any photons in the visible region and thus will be colorless. Substances with E_g values less than 1.7 eV can absorb all of them and thus will appear black. Substances with intermediate values of E_g will display a range of colors. For example, diamond (C) with an E_g of 5.4 eV is colorless, cinnabar (HgS) with an E_g of 2.0 is red, and galena (PbS) with an E_g of 0.4 is black [31, p. 170].

Figure 3.10 illustrates the difference in the spectra of substances undergoing ordinary charge transfer transitions and semiconductors. An example of the former, potassium permanganate, has a transition energy of around 2.3 eV (540 nm) and absorbs in the green region of the spectrum; the red and blue portions of white light remain largely unabsorbed, and so the color of permanganate is purple, or magenta, the complementary color to green. We observe an absorption maximum at 540 nm in the permanganate spectrum. On the other hand, cadmium sulfide, with a transition energy of around 2.4 eV (520 nm), and also absorbing in the green region, is a semiconductor which will absorb all photons with energies of 2.4 eV or greater. In other words, cadmium sulfide will absorb the entire blue-violet region of the spectrum, as well as the green, and reflect only the red–orange-yellow portion of the spectrum, thus appearing yellow. We observe in this case an

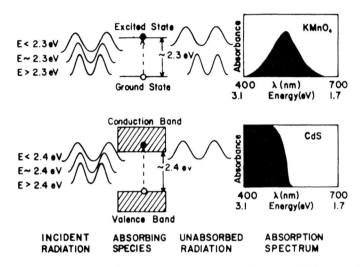

Fig. 3.10 Absorption spectra of potassium permanganate (above, exhibiting an absorption maximum at 2.3 eV) and cadmium sulfide (below, exhibiting an absorption edge at 2.4 eV) © 1980, American Chemical Society, Ref. [43]

absorption *edge* at 520 nm since only photons with wavelengths longer than that will remain unabsorbed [41, 42].

A good deal of chemistry is involved in understanding the color phenomenon. This discussion was necessarily limited to only a few cases of color production by atoms and molecules. Although human use of color dates back to the very beginnings of pre-history, it is only within the past century that the fundamental reasons for the existence of color have begun to be understood. In the following chapter, we will examine color usage from "the beginning," but will refer to much of what was discussed here.

References

1. Elschner C (1907) On the occurrence of silicate gems and other rare minerals in the Hawaiian Islands. Chem-Ztg 30:1119
2. Petursdottir SK, Bjornsdottir SH, Hreggvidsson GO, Hjorleifsdottir S, Kristjansson JK (2009) Analysis of the unique geothermal microbial ecosystem of the Blue Lagoon. FEMS Microbiol Ecol 70(3):425–432
3. Dalton J (1803) On the absorption of gases by water and other liquids. In Alembic Club Reprint No. 2 (1893) Foundations of the atomic theory. William F. Clay, Edinburgh, p. 25
4. See for example Lee Buescher's website: http://atomictimeline.net/index.php and Barcodes http://www.barcodesinc.com/articles/timeline-on-atomic-structure.htm. Accessed 11 Dec 2011
5. Mendeleev D (1869) On the relationship of the properties of the elements to their atomic weights. Zeitschrift für Chemie 12:405–406
6. Thomson JJ (1897) Cathode rays. Phil Mag 44:293–316

7. Goldstein E (1898) Über eine noch nicht untersuchte Strahlungsform an der Kathode inducirter Entladungen. Annalen der Physik 300:38–48
8. Rutherford E (1911) The scattering of the α and β rays and the structure of the atom. Proc Manch Lit Phil Soc, IV 55:18–20
9. Planck M (1922) 1920 Nobel prize address. In: Moulton FR, Schifferes JJ (eds) Autobiography of science 1950. Doubleday, New York
10. Einstein A (1905) Über einen die Erzeugung und Verwandlung des Lichtes betreffenden heuristischen Gesichtspunkt. Annalen der Physik 17:132–148
11. Moseley HGJ (1913) The high frequency spectra of the elements. Phil Mag 26:1024
12. Soddy F (1913) Intra-atomic charge. Nature 92:399–400
13. Chadwick J (1932) The existence of a neutron. Proc Roy Soc A 136:692–708
14. Bohr N (1913) On the constitution of atoms and molecules. Phil Mag 26:1–25
15. Balmer JJ (1885) Notiz über die Spektrallinien des Wasserstoffs. Ann Phys 25:80–87
16. Langmuir I (1919) The structure of atoms and the octet theory of valence. Proc Natl Acad Sci U S A 5:252–259
17. Russell C (1996) Edward Frankland: chemistry, controversy and conspiracy in victorian England. Cambridge University Press, Cambridge, p 108
18. Frankland E (1852) On a new series of organic bodies containing metals. Phil Trans 142:417–444; p 440
19. Kossel W (1916) 1. Über Molekülbildung als Frage des Atombaus. Ann Phys 49(IV):229–362; p 241
20. Lewis GN (1916) The atom and the molecule. J Am Chem Soc 38:762–786; pp. 777–778 (reprinted with permission, American Chemical Society © 1916)
21. Langmuir I (1919) The arrangement of electrons in atoms and molecules. J Am Chem Soc 41:868–934; p. 933 (reprinted with permission, American Chemical Society © 1919)
22. Pauling L (1931a) Quantum mechanics and the chemical bond. Phys Rev 37:1185–1186 (this introductory paper is not part of the series, but leads into it)
23. Pauling L (1931) The nature of the chemical bond. Application of results obtained from the quantum mechanics and from a theory of paramagnetic susceptibility to the structure of molecules. J Am Chem Soc 53:1367–1400
24. Pauling L (1931) The nature of the chemical bond. II. The one-electron bond and the three-electron bond. J Am Chem Soc 53:3225–3237
25. Pauling L (1932) The nature of the chemical bond. III. The transition from one extreme bond type to another. J Am Chem Soc 54:988–1003
26. Pauling L (1932) The nature of the chemical bond. IV. The energy of single bonds and the relative electronegativity of atoms. J Am Chem Soc 54:3570–3582
27. Pauling L, Wheland GW (1933) The nature of the chemical bond. V. The quantum-mechanical calculation of the resonance energy of benzene and naphthalene and the hydrocarbon free radicals. J Chem Phys 1:362–374
28. Pauling L, Sherman J (1933) The nature of the chemical bond. VI. The calculation from thermochemical data of the energy of resonance of molecules among several electronic structures. J Chem Phys 1:606–617
29. Pauling L, Sherman J (1933) The nature of the chemical bond. VII. The calculation of resonance energy in conjugated systems. J Chem Phys 1:679–686
30. Orna MV (1980) Chemistry and artists' colors: part I. Light and color. J Chem Educ 57:256–258
31. Nassau K (1983) The physics and chemistry of color. The fifteen causes of color. Wiley-Interscience, New York
32. Witt ON (1876) Zur Kenntniss des Baues und der Bildung färbender Kohlenstoffverbindungen. Ber 9:522–527
33. Orna MV (1978) The chemical origins of color. J Chem Educ 55:478–484
34. Griffiths J (1976a) Colour and constitution of organic molecules. Academic Press, New York, pp 82, 140

35. Griffiths J (1976b) Colour and constitution of organic molecules. Academic Press, New York, pp 17–53
36. Werner A (1893) Beiträge zur Konstitution anorganischer Verbindungen. Z Anorg Chem 3:267–330
37. Van Vleck JH (1932) Theory of the variations of paramagnetic anisotropy among different salts of the iron group. Phys Rev 41:208–215
38. Koren ZC (1996) Historico-chemical analysis of plant dyestuffs used in textiles from ancient Israel. In: Orna MV (ed) Archaeological chemistry: organic, inorganic and biochemical analysis. American Chemical Society, Washingon, pp 269–310
39. Przibram K (1923) Verfärbung und Lumineszenz durch Becquerelstrahlen. Zeit Physik 20:196–208
40. Rabinowitch E (1942) Electron transfer spectra and their photochemical effects. Rev Modern Phys 14:127
41. Yu PY, Cardona M (2004) Fundamentals of semiconductors: physics and materials properties. Springer, New York
42. Turley J (2002) The essential guide to semiconductors. Prentice Hall, New York
43. Orna MV (1980) Chemistry and artists' colors: Part II. Structural features of colored compounds. J Chem Educ 57:264–267

Chapter 4
Colorant Usage from Antiquity to the Perkin Era

4.1 Body Paint, Face Paint, Hair Coloring, Cosmetics

Images of butterflies, cats, dogs, fairies, ghosts, witches, wizards… kids of all ages love having their faces painted in these fanciful ways. They stand in a long tradition. From ancient times, body and face paints have been used for cosmetic, ceremonial, military, and religious reasons. There is reliable archaeological evidence that human beings have painted faces and bodies since the very beginning. The ancient Picts used red ochre, as well as woad (an indigo-bearing native plant). Julius Caesar remarked in book five of *De Bello Gallico*, "All Britons paint themselves with woad, which grows wild and produces a blue dye. This gives them a terrifying appearance in battle" [1]. Use of red ochre as body paint among Native Americans and the Maori people of New Zealand also appears to have been common. It was also used as a rouge or lip gloss for women in ancient Egypt [2]. Application of color to various parts of the body did not neglect hair. Ovid in the first century reported that the early Teutons colored their gray hair black with woad [3]. Henna enjoyed and continues to enjoy wide cosmetic usage in the henna-growing regions of the world (Asia Minor, the Middle East, and Australasia). The leaves of the plant, *Lawsonia inermis*, contain the colorant lawsone (2-hydroxy-1,4-naphthalenedione), a compound which has a strong affinity for protein and binds with it irreversibly [4]. Henna also has an unexpected health benefit as a natural sunscreen [5], but is not to be confused with "black henna," a mixture often containing p-phenylenediamine, which can cause severe allergic reactions [6].

Though beauty is said to be but skin deep, application of some cosmetics to that skin could be very hazardous indeed. Pigments for cosmetic use were very well-developed even five thousand years ago. In Egypt, eyelids were painted with kohl, powdered black galena (lead sulfide, PbS) to give that "Theda Bara" look, augmented by green malachite (basic copper(II) carbonate, $CuCO_3Cu(OH)_2$) [7, 8]. Cosmetic fashions followed changing ideals of beauty down through the ages, often in ignorance of, or by simply ignoring, mounting evidence of the toxicity of certain colorants [9]. For example, use of white lead [basic lead carbonate,

M. V. Orna, *The Chemical History of Color*, SpringerBriefs in History of Chemistry, DOI: 10.1007/978-3-642-32642-4_4, © The Author(s) 2013

$(PbCO_3)_2 \cdot Pb(OH)_2$], a staple cosmetic in ancient Rome, reached its apogee in the fifteenth through eighteenth centuries when the "dead white" look was all the go. Unfortunately, continued usage changed the word "dead" from its figurative use to its actual effect. Cosmetics of that era were also packed with a host of other deadly "delights:" mercury, arsenic, and phenol, among them. In 1936, Ruth deForest Lamb (1896–1978), Chief Educational Officer of the United States Food and Drug Administration (FDA), by publishing her book, *American Chamber of Horrors* [10], a dramatic, startling presentation of the problem of food and drug control and the serious dangers of many products available to the public, fired the opening volley in the fight to regulate cosmetics as well as food and drugs. A more recent work [11] explores this battle for safe cosmetics in the United States leading up to the Food, Drug, and Cosmetic Act of 1938. The author lays the background by reviewing the original 1906 Food, Drink, and Drug legislation, its advocates, and its shortcomings, among which was the omission of cosmetics from FDA regulation, not as an oversight but as one of many concessions the framers intended to be rectified through amendment. During the 1920s and 1930s a phenomenal increase in cosmetic use, and its dangers, intensified the need for government supervision.

Unfortunately, the needed supervision is mostly by hindsight and after the fact. In 2007, an independent watchdog group found that one-third of the lipsticks on the market contained dangerous levels of lead. There is no lead limitation spelled out by the FDA. "The U.S. Food and Drug Administration does oversee cosmetics, but it's an after-the-fact kind of oversight, where unsafe products can be taken off the market once they have been proved to be unsafe. Unlike drugs, cosmetics don't have to go through clinical trials before they go on the market", declares author Teresa Riordan [12]. Does she or does not she have lead in her lipstick? Only time will tell.

4.2 Ceramics, Glasses, Glazes and Stained Glass

Ceramic materials constitute some of the earliest artifacts of humankind. This is not surprising since ceramics are made of the stuff of the earth, the clay-like material that can be found almost anywhere. It is even less surprising when we consider that all ceramic materials are noted, once fired, for their durability, hardness and resistance to attack by heat and corrosive substances. And even less surprising when we realize that ceramic materials constitute the major finds in almost all archaeological digs, even the most ancient—apart from stone-working, pottery manufacture is the oldest of all manufacturing techniques.

Color enters the picture when we consider that ceramic materials, when unglazed, are quite porous: a rough, unglazed flower pot is a modern example. When clay, or kaolin, from which a ceramic piece is formed, is fired, it goes through a series of changes as the temperature of the oven or kiln is raised. If the temperature is high enough (1,000–1,250 °C), mullite ($Al_6Si_2O_{13}$) and glass form from the feldspar and silica present in the clay body; this process is known as vitrification and serves to seal the entire structure into a hard, strong mass [13]. The porosity of an object is inversely proportional to the degree of vitrification, and no ceramic article, consisting primarily of kaolin platelets, ever becomes 100 % vitrified. Hence it is necessary to coat fired pottery with a glaze that performs three functions: it makes the pottery nonporous and watertight; it enhances the strength of the object; it adds beauty to the object.

4.2.1 The Nature of Glasses

A ceramic glaze is a thin glass-like layer applied to a pottery surface and then fired in a kiln. Since the chemistry involved in glaze formation differs only slightly from that used in the manufacture of glass, it would be useful to examine the nature of glasses first.

What is the difference between a ceramic material and a glass? Initially, one can outline their differences by looking at their properties. Ceramics are shaped at room temperature, while glasses are shaped at elevated temperatures. Ceramics harden with application of heat; glasses harden upon cooling. Ceramics are composed of silicates and aluminates mixed with silica and some salts in minor quantities; glasses are largely silica mixed with fluxing agents which lower their melting point. Ceramics are almost visibly porous; glasses are nonporous. Ceramics are microcrystalline in structure; glasses are noncrystalline in structure. Both are nonconductors of heat and electricity.

Pure silica has a melting point of around 1,700 °C, a rather unattainable temperature with the means available in ancient and medieval times, and as a liquid, it is too stiff and viscous to work with even at any reasonable temperature. Thus, both melting point and viscosity must be reduced by adding fluxing agents, such as sodium and calcium salts. These agents are able to lower the melting points of crystalline materials because at elevated temperatures they undergo chemical reaction with the silica to generate alkali silicates, which melt at lower temperatures than the pure silica. These fluxes are then converted to their corresponding oxides during the glass-forming process and are bound via ionic bonds to the silicate network that makes up the bulk of the glass body.

4.2.2 The Desired Properties of Glazes: Colorants in Glazes

A glaze is formulated in order to impart beauty to an object, to reduce porosity, and also to increase its overall strength and chemical resistance. So it must be fluid enough to fill the external pores of the ceramic piece upon firing, but viscous enough not to run off the piece in the process. Furthermore, it must have the capacity to be fired at higher temperatures than glass and possess an affinity for the clay body. To achieve this, a glaze must contain a higher proportion of alumina than glass, which has a high (~ 70 %) silica content. Widespread use of glaze on pottery first occurred in Mesopotamia after 2000 BCE [7, 8].

Early potters discovered that the compounds of copper and iron were ideal colorants for glazes. Copper compounds produced turquoise blue and green colors; iron compounds produced yellow, green, and brown colors. Progress in finding additional colorants was slow. Copper and iron compounds remained the mainstay of ceramic colorants until around the year 1200 CE. Around this time, Chinese potters introduced red copper(I) oxide, Cu_2O, and later, yellow lead antimonate, $Pb_3(SbO_4)_2$, blue cobalt silicate, Co_2SiO_4, and manganese silicate, Mn_2SiO_4, which produces a purple-brown color. The Moors introduced tin(IV) oxide as a white colorant, and colloidal gold was known in Europe as a pink colorant from the latter part of the sixteenth century.

An interesting colorant known from ancient times and still in use today as a high-grade glaze colorant of unsurpassed clarity and brightness is "Purple of Cassius". It is produced by adding a solution of tin(II) chloride to a very dilute solution of gold chloride, $AuCl_3$, producing a precipitate of hydrated tin(IV) oxide interspersed with finely divided elemental gold, the reduction product. The gold colors the tin oxide precipitate brown, purple, or red, depending upon the original concentration of the solution [14].

As chemistry replaced alchemy in the eighteenth century, progress in glaze colorants became more rapid. Chromium salts were introduced in France, and their use quickly spread to the rest of Europe. As more new materials became available, pottery manufacturers began to produce their own colorants and colorant formulations, which often took the form of "secret formula" books. Over the following century, glaze formulation gradually became a manufacturing specialty in its own right, and by the end of World War II, in-house manufacture of colorants by pottery makers was a thing of the past.

The amateur potter can produce the complete spectrum of glaze colors by the use of oxides or carbonates of only eight metals (chromium, cobalt, copper, iron, manganese, nickel, titanium, vanadium). Numerous patents have been issued involving other elements such as zirconium, antimony, cadmium, gold, selenium, and uranium. The formation of a glaze colorant in the firing process is not simple, and the study of phase diagrams at high temperatures and under various firing conditions has resulted in postulated mechanisms of color formation [15].

This brief discussion of glaze chemistry makes it all too clear that glaze formation is one of the more complex types of high-temperature chemical reactions. Many different reactions, in different phases, and under different firing conditions, may take place. However, it is essential that the glaze be stoichiometrically compounded in such a way that it fulfills exactly the required ultimate melted analysis, i.e., with no starting material left over and unreacted. Such compositions, while the results of trial and error before the advent of chemistry, may now be precalculated on the basis of the series of chemical reactions which take place in the firing process. This is one reason why studies of the exact mechanism of glaze formation are so important.

4.2.3 Stained and Painted Glass

Despite Pliny the Elder's (23–79 CE) whimsical account of the origins of glass in his *Historia Naturalis*, modern scholarship places Mesopotamia and Syria as the countries where glass was first made. It was later introduced into Egypt during the reign of Tuthmosis III (1480–1425 BCE), after which Egypt became the center of glass production. Centuries later, two great technological advances led to the widespread use of glass for a variety of purposes. First, the invention of the blowpipe, often dated from between 300–20 BCE and introduced into the Roman world in the first century CE, allowed for the manipulation and eventual mass-production of glass objects. Secondly, the introduction of grooved strips of strong, yet malleable, lead as a matrix for glass pieces (lead cames) allowed for the widespread use of glass in an architectural context. Although this crucial technological advance cannot be dated, fragments of glass in lead have been found in Roman ruins at Trier in Germany and Senlis in France. Household windows came to be glazed during the heyday of the Roman Empire, but it was the use of both stained and painted glass in Christian churches that secured its future as a decorative art form from as early as the fourth century CE. In fact, not only did stained glass play an aesthetic role in churches and cathedrals, but it also was a major form of religious instruction in a largely illiterate society [16].

It is not known when paint was first applied and fired onto window glass. This development is important because it is the combination of colored glass and painted detail that has become characteristic of stained glass. The most valuable source of information on the making of a medieval window is the treatise by the German monk Theophilus (Roger of Helmarshausen) [17] who guides the craftsman through the design process to the final mounting of the window. After the cutting of the glass, the most skillful operation of all was the application of the paint, made of iron, copper filings, and copper oxide, mixed with ground glass and a liquid binder to help the mix adhere to the glass surface. This paint gave a gray-black to brown stain, depending upon concentration; the only other early alternative to the medieval artist was yellow or silver stain (silver nitrate or silver oxide) introduced around 1300.

Although the role of the lead cames was to support and hold the glass pieces together, they were often an integral part of the overall design, a fact often obscured by later repairs. In addition, lead deteriorates over time and has a very low resistance to stress compared with other metals, as well. This results in lead deformation since it is under the strain of supporting the weight of the glass [18]. For this reason, stained glass windows must be observed very carefully for possible buckling and removed and re-leaded if called for, but the frequency of the conservation effort must be done on a case by case basis. Otherwise, the window is at risk of succumbing to gravity, with the glass pieces falling out of their frame. An example of this unfortunate event and even more unfortunate restoration can be seen in Fig. 4.1. This figure also illustrates the appearance of silver stain very well.

Fig. 4.1 Stained and Painted
Glass (ca. 1200 CE) in the
Narthex of Cartmel Priory,
Cumbria, England.
Photograph by M. V. Orna

An additional note on lead cames in a modern context is important. Lead is now extruded into cames by precisely measured machinery that can cut down greatly on the amount of lead used, and hence on the weight of the cames. Furthermore, addition of certain metals or metalloids, such as antimony, to the lead structure serves to strengthen the lead and give it a life exposed to the weather for perhaps many centuries. Stained glass windows can be further protected by the preventive conservation technique of protective isothermal glazing.

Although the colors available for stained glass were somewhat plentiful, as outlined in Sect. 4.2.2 above, and evidenced in the windows of the great medieval cathedrals, the palette of the glass painter considerably expanded beyond silver stain in the sixteenth century with the addition of a new range of vitreous enamel pigments. The earliest colored stain was a red pigment called "Cousin's red" after its supposed inventor. Introduction of brilliant blues, greens, and purples followed, and were immediately taken advantage of in the construction of King's College Cambridge, a prime example of stained and painted techniques [16].

The decades between 1870 and 1930 saw the proliferation of stained and painted glass in the United States. Two major glassmakers, John LaFarge (1835–1910) and Louis Comfort Tiffany (1848–1933), are practically household words as a result. They drew their inspiration largely from their European predecessors, but departed from traditional historic revivals and geometric patterns then in vogue. Other glassmaking studios brought their own family traditions from the Old World, and continue today to build upon them while at the same time taking bold, innovative steps. One such studio is that of Emil Frei of Saint Louis, originally from Bavaria. An example of one of their traditional works is shown in Fig. 4.2. Their website shows the complete range of their work [19].

Fig. 4.2 "Pietà" executed by Emil Frei and Company, Saint Louis, Missouri, 1924. Chapel of the National Shrine of Our Lady of Prompt Succor, New Orleans, Louisiana

4.3 Artists' Colorants

What is the difference between a pigment and a dye? According to the principles of color technology, pigments are applied to a surface with the use of a vehicle that allows it to be spread; dyes need no such binder and can inhere in the substrate directly. Thus, any substance may be classified as a pigment or a dye depending upon how it is applied to the substrate, and both should be discussed under the more general classification of "colorants".

For a pigment to be useful, it must be insoluble in the binding medium so as to form a suspension. Otherwise it may "bleed" into successive layers of paint and render itself out of the control of the workman or artist. For a dye to be useful, it must be soluble in a dyebath or at least able to be rendered soluble prior to use. These distinctions and their technical implications will be discussed later in this section.

4.3.1 Literature Sources for Ancient, Medieval, and Renaissance Technical Practice

Our present knowledge of colorant usage dates back to Pliny the Elder's "Natural History," Book XXXV, which consists of a massive 59 chapters describing all that was known in the ancient world about the art of painting. This work, practically the only one that describes the methods and materials of artists of the time, is used as a reference for the history of art even today. In particular, Chap. 12 describes pigments not derived from metals and artificial colors; Chap. 14 through 30 describe individual pigments, including Egyptian earth and ochre; Chap. 31 discusses colors that do not admit of being laid on a wet coating; Chap. 32 describes what colors were used by the ancients in painting [20].

In the third century CE, two Greek papyri containing technical information were written in Alexandria. Now residing in Leyden and Stockholm, they are the last surviving technical manuscripts written in a European language until the reappearance of the technical tradition around the year 800 [21, 22]. These manuscripts deal with the manufacture of metals, artificial gems, alloys, and colorants, and somehow the tradition they represent survived over five centuries only to re-emerge in two Latin manuscripts in the late eighth or early ninth century. One of these, part of the *Liber Pontificalis* in the Codex Lucensis 490 [23] at Lucca (in Tuscany, Italy) deals with recipes for making dyes, pigments, metals, and related crafts. The second manuscript, contemporary with the Lucca, is a compilation of virtually everything contained in the Lucca, plus many additional recipes, and was apparently translated north of the Alps. This is called the *Mappae Clavicula*, or *A Little Key to the World of Medieval Techniques*, and survives in a fairly complete twelfth century document [24]. Together with these two manuscripts, three others, the tenth century work of Heraclius, *De diversibus artibus Romanorum* [25], the twelfth century treatise of Theophilus [17], and the fifteenth century *Il Libro dell'Arte* of Cennino Cennini [26] form the foundation of our knowledge of the medieval and Renaissance technology of painting and craftsmanship. The latter, often translated as *The Craftsman's Handbook*, is a "how to" on Renaissance art, containing information on colorants and painting techniques, including oil painting and fresco.

These primary sources were digested, translated, and embellished in a number of secondary sources that form a body of literature indispensable to the chemist interested in the history of colorants as applied to the arts. They are listed here in order in which they were first published. Mary Philadelphia Merrifield's 1846 study on fresco painting [27] as practiced by the classical Italian and Spanish artists was translated, as she ingenuously describes, by her two young sons in a quite literal manner. The preliminary section is an inquiry into the nature of colors and pigments used by these schools, separated by color types. In the main part of the study, the author draws on the observations and directions of the principal authors who treated the subject between 1000 and 1779 A.D. (such as Theophilus, Cennini, Alberti, Vasari, Guevara, Borghini, Armenino, Pozzo, etc). Sir Charles

Lock Eastlake's great work [28] comes to us in two volumes. Volume one, first published in 1847, traces the recorded pattern of oil painting from its invention. Heraclius and Theophilus are the author's chief sources for describing the earliest practice of oil painting. From there the author describes fourteenth century oil painting, fresco and wax painting, the introduction of "improved" oil painting by the Van Eycks in 1410 (although the technique was already well-documented by Theophilus [17]), and all that followed with respect to colorant preparation and oil preparation in the Flemish school. The second volume, published posthumously in 1868 from fairly complete preparatory documents, recapitulates the characteristics of the early Flemish school and then completes the work with a history of the Italian school. Since its original publication in 1849 (J. Murray, London), Merrifield's two-volume work on the technology of medieval and Renaissance oil painting [29] has been one of the foremost among a scarce handful of valued reference books dealing with the subject. The work reprints (with the original language version and its English translation on facing pages) manuscript collections on painting and related arts dating roughly from the twelfth through the seventeenth centuries. The manuscript describes oil painting practices in several Italian cities, and in France and Brussels. Most of them are recipe books, revealing the artists' methods of making, purifying, grinding, and dissolving (actually suspending) many different kinds of pigments; of preparing wood and cloth for painting; of making inks, dyes, and glues; and much more. Although oil painting receives the primary emphasis, the treatises also cover the processes involved in making miniature paintings, mosaics, and paintings on glass, as well as what is entailed in the craft of gilding, glazing, cutting precious stones, and many others.

Daniel V. Thompson's *The Materials of Medieval Painting* [30], first published in 1936, is a work based on years of study of medieval manuscripts and enlarged by laboratory analysis of medieval paintings. It is an invaluable resource for art historians, students of medieval painting and civilization, and historians of culture. It discusses carriers and grounds, binding media, pigments and other coloring materials, and metals used in painting. Gettens and Stout's *Painting Materials: A Short Encyclopedia* [31, 32], originally published in 1942, is a classic in the varied activities of painting and conservation, cultural research, chemistry, physics, and paint technology. The book contains over 100 entries on pigments and inert materials from alizarin to zinnober green, complete with references to primary sources when available, and the coverage is exhaustive. Rosamund D. Harley's *Artists' Pigments c.* 1600–1835 [33] is a monumental little book that fills a big gap, namely the history of materials of English painting gleaned from documentary sources little known and almost unavailable to the average investigator. After a description of the literary sources used and the sources for the history of the color trade, the author devotes the following nine chapters to colorants organized by color and nature: inorganic blues, organic blues, greens, inorganic yellows, organic yellows, inorganic reds and purples, organic reds, browns, blacks and grays, and finally whites. A final chapter relates the growing availability of artists' pigments to progress in the sciences, particularly chemistry. An appendix entitled "Patents for Colour-Making in the Early Seventeenth Century" documents the reasons why industrial activity in that century never reached the proportions of the eighteenth century industrial revolution.

No literature listing on this topic would be complete without including the three volume handbook on artists' pigments published by the National Gallery of Art as a continuation of a series of monographs originally published in *Studies in Conservation* between 1966 and 1974. The pigments chosen were those that have played a major role in the history of painting and much of the information that the authors supply is based on original laboratory and bibliographic research. Volume 1 [34] contains monographs on yellows (Indian, cobalt, cadmium, chrome, lead antimonate), whites (barium sulfate and zinc white), reds (red lead, minium, and carmine), and green earth. Volume 2 [35] treats blues and greens (azurite, blue verditer, natural and artificial ultramarine, smalt, verdigris, copper resinate, malachite green, and green verditer) as well as lead white, lead–tin yellow, vermilion and cinnabar, and calcium carbonate whites. Volume 3 [36] covers more blues and greens (Egyptian blue, indigo, woad, Prussian blue, emerald green, Scheele's green and chromium oxide greens), yellows and browns (orpiment, realgar, gamboge, Vandyke brown), as well as madder, alizarin, and titanium dioxide whites. Each monograph begins with an extensive history of the pigment, including terminology, and continues with properties, chemical composition, sources, preparation, identification and characterization, and notable occurrences from ancient to recent times.

While the pigment handbook concentrates on natural inorganic colorants normally used as pigments, several in the list are also used as dyes. Dye handbooks that chronicle dye usage from ancient times also abound. The most outstanding volume in this class is Brunello's classic, "The Art of Dyeing in the History of Mankind," [37], first published in Italian in 1968 and subsequently in English translation in 1973. Some remarks from the preface are worth quoting here:

> [People] in the thousands of years of their journey have experimented with numberless materials to produce the most beautiful and lively colorings: first on the human body, later on clothing. From prehistory on, men of different races and peoples have engaged in this technological enterprise, starting from the elementary applications of colored earth to arrive…at the synthesis of artificial dyestuffs. Through Chemistry, one of the highest expressions of the human intellect, man has been able to surpass even Nature itself in the creation of new substances, which, in their turn, have given rise to a veritable symphony of colors.

The author, in the next 467 pages, follows the evolution of this journey from using the natural materials available in various localities, to Perkin's breakthrough of 1856: in prehistory, in ancient non-European civilizations, in the classical period, in the middle ages, the Renaissance, and the centuries leading up to 1856.

4.3.2 Pigment Use in Manuscript Painting

Although the earliest surviving manuscripts from the Roman and Byzantine empires date to around the fifth century CE, the vast majority are from the tenth century onward until manuscript production died out with the invention of the printing press. For the most part, illuminated manuscripts, that is, those that contained painted illustrations, were of a religious nature and the product of monasteries. It was actually the continuous production of these manuscripts that preserved the language, culture and literature of ancient Greece and Rome.

The pigments used in these manuscripts were produced from mineral and plant sources and applied to prepared sheets of parchment or vellum using a binder such as egg yolk mixed with other substances such as wax or even urine. There is a body of literature on pigment analysis of Armenian and Byzantine manuscripts [39–42] that indicates that the Armenian palette relied heavily on mineral pigments, whereas the Byzantine palette was found to consist primarily of organic pigments. Other important observations were:

- Natural ultramarine as a blue pigment and vermilion (natural mercury(II) sulfide) as a red pigment are common to all of the manuscripts examined;
- Azurite (blue basic copper(II) carbonate) was detected, but only in manuscripts later than the fourteenth century, when it may have been imported from the West;
- No true green pigment was found in any of the manuscripts; the green hue was achieved by mixing blue and yellow pigments;
- The purple hues were obtained by mixing varying amounts of vermilion or madder with blue pigments;
- The yellow pigment, orpiment (arsenic(III) sulfide), was used extensively in Armenian manuscripts but was absent from the vast majority of Byzantine manuscripts.

Additional data for Persian, Turkish, Indian and Iranian manuscripts for purposes of comparison can be found in Ref. [43]. This information has been valuable to the art historian for understanding more about the artistic process itself and for tracing lines of influence from one culture to the next.

Analysis has also uncovered at least one forgery. The date of fabrication of Ms 972 from the University of Chicago Special Collections, familiarly known as the "Archaic Mark," had been tentatively attributed to the twelfth century. Analysis showed that an iron blue [42] was ubiquitous in this manuscript, raising doubts about its authenticity. The iron blues are the first of the artificial pigments with a known history and an established date of first preparation. The color was made by the Berlin color makers Johann Jacob Diesbach and Johann Konrad Dippel (1673–1734) in or around 1706 [44, 45]. Moreover, according to Gettens and Stout [31, 32], the material is so complex in composition and method of manufacture that there is practically no possibility that it was invented in other times and places. This fact, in addition to other evidence [46]—radiocarbon dating of the parchment and the discovery of additional "anachronous materials" such as blanc fixe (barium sulfate) and cellulose nitrate—suggest that the Archaic Mark originated some time much later than its purported twelfth century fabrication, more than likely the early twentieth century.

4.3.3 A Digression on Two Blue Pigments: Iron Blue and Ultramarine

These two pigments, found in manuscripts and in paintings, have such interesting chemistry and such a legendary history that they deserve special mention at this point.

4.3.3.1 The "Iron Blue" Pigments

The so-called "Iron Blue" pigments are a special case of charge transfer (please see Sect. 3.5.3), namely between two atoms of the same element in two different oxidation states. The two possible oxidation states of Fe(II) and Fe(III) in compounds are present in the iron blues; such a compound is called a mixed-valence compound. The history of mixed-valence chemistry is almost as old as the history of chemistry itself. It actually begins in 1706 with the accidental synthesis of one of the first of these compounds, Prussian blue, with the formula $KFe(II)-Fe(III)(CN)_6$. Prussian blue owes its intensely blue color neither to ligand-field transitions, nor to charge-transfer transitions of the type metal $\rightarrow$ ligand or ligand $\rightarrow$ metal, but to an intervalence transition of the type $Fe^{II} \rightarrow Fe^{III}$. This sort of transition is highly probable (quantum mechanically "allowed") and often so strong that it dominates the visible spectrum and is responsible for the color of the substance [47].

The iron blues have a number of different formulas depending on the method of manufacture. For example, two other formulas are $NH_4Fe(II)Fe(III)(CN)_6$ and $Fe(III)_4[Fe(II)(CN)_6]_3xH_2O$. The different iron blues have different physical properties such as ease of grinding and transparency, and consequently find numerous uses. In their pure forms they are sometimes called Prussian blue, Berlin blue, Milori blue, Chinese blue, Paris blue, etc. In their less pure forms, they are variously called Antwerp blue, Brunswick blue, Erlanger blue, Hamburg blue, and laundry blue. All forms are designated by the single *Colour Index* (C.I.) number Pigment Blue 27 [48, 49]. (The *Colour Index* is the response of the color and dye industry to the need to classify colors following the explosive syntheses of colored compounds in the mid-nineteenth century.) Use of Prussian blue by artists spread rapidly and has been identified in paintings by artists as disparate as Canaletto, Watteau and Hogarth [50].

Fig. 4.3 "Prussian Blue" by Thomas Phillips, 1816. Courtesy of Linda and Daniel Bader, Milwaukee, WI. Reproduced with permission of the owners

An interesting art-historical sidelight: It is well-known that Michael Faraday (1791–1867) and his teacher, W.T. Brande (1788–1866) used Prussian blue synthesis as the star attraction in their lecture demonstrations at the Royal Institution and elsewhere. Alfred Bader, a chemist-collector with a sleuth-like bent in matters of art, came into possession of a painting (Fig. 4.3) in 1989 and immediately identified the activity taking place as synthesis of Prussian blue. It took a great deal more sleuthing for him to be able to approximate the date of the painting (1816), identify the artist (Thomas Phillips), and finally to identify the figures as the as-yet-unknown twenty-four-year-old Faraday with the twenty-eight-year-old Brande—circumstantial evidence that this is the earliest portrait of the great and good scientist, Michael Faraday [51]!

4.3.3.2 Ultramarine Blue

Natural ultramarine blue was one of the most expensive and desirable of the natural blue pigments. Cennino Cennini says: "Ultramarine blue is a color illustrious, beautiful, and most perfect, beyond all other colors; one could not say anything about it, or do anything with it, that its quality would not still surpass" [26]. It was called "ultramarine" because it came from "beyond the sea," namely the Caspian Sea, because its only known source until the nineteenth century was the near-inaccessible caves of Badakshan in Afghanistan. In 1271 Marco Polo described his visit to these caves and subsequently, the mineral was imported to Venice, where it more than likely was also processed. The parent mineral is called "lapis lazuli", or simply "lapis" to artists, and the purified mineral extracted from it is termed lazurite. It is chemically a sodium calcium aluminosulfosilicate of varying composition. Simple grinding and washing is sufficient for very pure forms of the mineral, which are rare. The normal method of extraction is to mix the ground mineral into a ball of waxy material and then knead the ball under a lye

Fig. 4.4 Examples of good
quality (*left*) and poor quality
(*right*) ultramarine.
Photographs by M. V. Orna

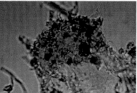

solution (usually potassium carbonate leached from wood ash). Presumably because of preferential wetting, the large blue particles of the mineral settle to the bottom of the solution during kneading and can be separated out from the colorless material. Several fractions can be extracted, yielding pigment of successively lesser quality. Merrifield [29] and Thompson [30] speak of contracts for its inclusion in commissioned works because of its great price, often more than the price of gold.

Figure 4.4 compares two photomicrographs of ultramarine blue. On the left is fairly high quality ultramarine, as can be seen from the large amount of blue pigment in the colorless matrix; on the right is a sample of "ultramarine ash," a low quality ultramarine with a grayish color. The pigment itself is almost completely lightfast and impervious to alkali, but it will decompose readily when treated with acid. Although the origin of its color has been investigated extensively, it is still unclear. Since it contains the highly reactive polysulfide radical anions, S_2^- and S_3^-, it has been postulated that the absorption band is due to mixed-valence charge transfer within a clathrate matrix [52]. It has been found in Armenian illuminated manuscripts [39, 43], as well as in paintings by Botticelli, Rubens, Van Dyck, Titian and Poussin, among others. When French chemist Jean-Baptiste Guimet (1795–1871) succeeded in synthesizing artificial ultramarine in 1828, the bottom fell out of the market for the natural material: the synthetic pigment cost less than one-tenth the price of the natural. Actually, Guimet asserted that he had succeeded in this synthesis two years earlier, but had not published his results. Almost simultaneously and independently, Christian Gmelin of Tübingen and F.A. Köttig of Meissen perfected processes of the same kind. However, it came to be known as "French ultramarine" possibly because of its discovery and long production in France. Artificial ultramarine, in contrast with natural ultramarine, is finely divided and homogeneous [31, 32]. As one might expect, examples of the use of the artificial product can be found in the works of many of the nineteenth century French Impressionists [53]. However, the search for the definitive reason for the blue color still continues.

4.3.4 Pigment Use in Mosaic Painting

Thompson [30] places the development of fresco painting within the tradition of mosaic painting, which was an Italian tradition up until the thirteenth century. Given the fact that each tiny *tessera* of the mosaic piece had to be cemented into place a small portion at a time so that the cement would not dry until everything was in place, this was an extravagant method indeed. Nevertheless, we see its practice of using these tiny pieces of colored glass or stone throughout the ancient world. Mosaic *tesserae* utilize the same coloring materials used in stained glass unless they are simply cut from naturally colored stone or marble. *Tesserae* can be made of opaque glass fired in large slabs in a kiln with various pigments having been added to confer the color. Gold and silver tesserae are made by sandwiching real gold and silver leaf between two glass layers and firing them twice in a kiln to embed the metal.

The technique was adopted by the Christian church for the decoration of basilicas and many examples of this technique comprise part of the near-obligatory tourist route in Italy. The 20,000 or so tourists per day who visit Saint Peter's Basilica in Rome may not be aware of the fact that virtually the entire interior vaulting of this enormous building is composed of mosaics. This "trompe l'oeil" decorative program, begun in 1578, continues today through the artists of the Vatican Mosaic Studio who not only carry on an ancient tradition, but break new ground with modern stucco formulas and new applications of the filament enamel technique [54].

4.3.5 Pigment Use in Fresco Painting

Although Americans and Europeans are more familiar with the magnificent Renaissance frescoes that grace the pages of every art history textbook, fresco painting has been around for millennia. The technique is one of the most durable under the right conditions and if the associated building survives. Indeed frescoes remain from walls in Pompeii, from the Roman catacombs, from ancient public buildings, and from the Herodian era in ancient Israel. Archaeologists have been particularly diligent in the latter, and the colors used in Jericho and Masada have been analyzed and found to match those recommended by Vitruvius and Pliny [55]. Figure 4.5 shows a fresco fragment from a Herodian palace at Masada, Israel.

The colors have been analyzed and contain the following: yellow is limonite, $FeO(OH) \cdot nH_2O$; red is hematite, Fe_2O_3, mixed with some galena, PbS; black is soot, C, mixed with iron compounds; green is a mixture of Egyptian blue, $CaCuSi_4O_{10}$, and yellow ochre, $Fe_2O_3 \cdot H_2O$). During the Middle Ages and the Renaissance, especially in Italy, the technique was used universally to adorn churches, cathedrals and government buildings. Perhaps the most famous fresco of all is the monumental work done by Michelangelo on the Sistine Chapel vault from 1508–1512. Perhaps the most remarkable of the early modern frescoes are the Moldavian painted churches built approximately between 1487 and 1583. One of the greatest tourist attractions of northern Romania, the outdoor frescoes have remained bright and beautiful through the centuries because of their mineral pigment nature.

Fig. 4.5 Fresco from a
Herodian Palace at Masada,
Israel. Photograph by
M. V. Orna

Fresco painting, when done as *buon fresco*, that is, right on the wet plaster
ground, is one of the most demanding of all techniques because it is so unfor-
giving. It is actually a combination of painting accompanied by a chemical
reaction. The ground itself is prepared by first calcining (heating) calcium carbonate until it forms lime, calcium oxide. The lime is then slaked with water to
form calcium hydroxide, traditionally in lime pits where it was allowed to age for
as much as 10 years. The plaster ground is made by mixing the slaked lime with
sand or marble dust in varying proportions. Prior to painting, several layers of
plaster are laid down on the wall, each successive layer containing sand with
smaller particle sizes. Hardening of the plaster involves absorption of the water
into the wall, evaporation of water, and reaction of the lime with atmospheric
carbon dioxide to produce calcium carbonate prior to painting (carbonation).
When the wall is dry, the artist prepares the cartoon, or outline of what is to be
painted. Then the next and final layer of plaster, the *intonaco*, containing much less
sand (and often, marble dust) than the other layers, is laid on the wall. The outline
of the drawing is placed on the surface and traced into the wet plaster by incising
or by pushing a dark pigment through holes in the cartoon punched along the lines
of the drawing. Pigments are mixed with water and brushed onto the wet plaster;
they are fixed to the wall by the carbonation of the *intonaco* layer. Obviously, once
the layer becomes carbonated, which takes about seven hours, the fixing of the
pigment must cease—this is why an artist can only prepare as much *intonaco* as
can be worked in one day, called a *giornata*. Merrifield [27] emphasizes the need
for the pigments to be natural mineral colors and compatible with lime.

Because the porous mortar backing of a fresco allows for the transport of
dissolved salts to the back of the painting, deterioration can take place when these

salts crystallize and consequently expand, pushing their way through the painting to the surface. If the salts are sulfates, they will react with the calcium to form calcium sulfate (gypsum), which is more soluble than calcium carbonate. Dissolution and recrystallization of calcium sulfate with changing humidity can greatly damage a painting if the process occurs near the surface, so gradually substituting barium for calcium is now an accepted conservation method for damaged frescoes because barium sulfate is less soluble than the gypsum. However, the method is so slow that it sometimes takes years to restore a fresco—for this reason, the frescoes of the Brancacci Chapel in Florence were "under wraps" for nearly eight years.

4.3.6 Pigment Use in Oil Painting

Although some art historians place the emergence of oil painting in the early 15th century with the innovations of Jan and Hubert Van Eyck [28], oils were most certainly used long before that, but perhaps with a different purpose, as Thompson [29] suggests. He says that many works were "glazed" with oils after completion to lend a different tone to the painting or to change the color. However, oil painting proper, no matter when it began, involves the mixing of a pigment with a drying oil. We know that at about 1300 CE, the total list of pigments stood at some three dozen [56]. Then, from about the fifteenth century on, with the advent of more chemical know-how, many more were synthesized, and dates for when they entered the artist's palette are well-known and documented [31, 32]. The range of drying oils is far more limited than pigments; perhaps the most well-known of these is linseed oil. In order to be classified as a drying oil, the oil must have a sufficient number of double bonds in its triglyceride chains so that over time, the oil can polymerize into a film that can be said to be one giant molecule. Unless this happens, the painting will always be fluid and tacky. Another requirement for a paint is good hiding power unless the artist wishes the viewer to see what lies beneath the surface. To obtain an opaque paint, the oil (binder) must have a refractive index different from that of the pigment. Since the pigment is suspended in the binder, light will be unable to travel in a straight line, and the ground of the painting will be invisible. This is illustrated in Fig. 4.6 where we see an incident ray undergoing multiple reflections and refractions in the pigment particles, which are transparent. The ray may even be reflected back into the air. If the refractive indices of the binder and the pigment were very close to one another, then the light ray would travel through the paint in a nearly straight line, and be reflected from the surface of the ground in such a way that the image of the ground could be visible. Such a paint is transparent. One example of a transparent paint is calcium carbonate in linseed oil. Calcium carbonate has a refractive index of 1.50–1.64 and linseed oil's refractive index is 1.484. The difference between the two is so small that this makes for a transparent paint. On the other hand, water's refractive index is 1.33. If calcium carbonate were suspended in water, the refractive index difference is about 0.17, a considerable difference, and so the paint is opaque.

Fig. 4.6 Pathways of an incident ray through a paint where the refractive indices of the binder and of the pigment are different. © 1998, M. V. Orna

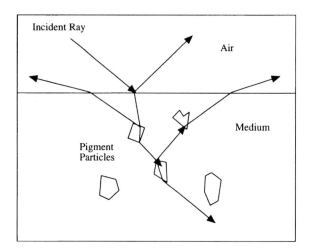

However, when the water dries, then we are looking at a refractive index difference of 0.33 between that of air (1.00) and calcium carbonate. This explains why whitewash is so effective.

One of the more durable drying oils that has been in use for centuries is linseed oil, pressed out of the seeds of the flax plant—a scarce commodity until the linen industry expanded and thus made more flax available for other purposes as well. Hardening of the soft linseed oil films by rosin and adding turpentine created an enhanced varnish for wooden floors, fine furniture and for Stradivarius violins.

There are other factors that need to be considered to make up a good paint, including pigment particle size, solubility of the pigment in the medium, media adsorption, reactivity with the medium, etc. There is so much more to color than meets the eye!

4.4 Dyes

While many techniques existed and were well-documented, the versatility of dye-stuffs was limited by availability and in some instances to the difficulty of obtaining the colorant. Apparently when God spoke, the ancient Hebrews did as commanded despite these difficulties. For example, from the Book of NUMBERS 15:38 we read: "Speak unto the children of Israel, and bid them that they make them fringes in the borders of their garments throughout their generations, and that they put upon the fringe of each border a cord of blue," and in Exodus 39:1 we read: "And of the blue, and purple, and scarlet, they made finely wrought garments, for ministering in the holy place, and made the holy garments for Aaron; as God commanded Moses". In Chap. 1, we saw that carmine cochineal coming from the New World was a very important Spanish import to the point where cochineal-laden galleons were the prime targets of pirates on the high seas. It was also versatile if treated properly. For

Fig. 4.7 100 mg of Tyrian
Purple, 6,6′-Dibromoindigo
(equivalent to the dye from
1,000 snails). Synthesized 15
December 1992. Courtesy
of Jan Kochansky

example, Cornelis Drebbel's accidental discovery that a tin "salt" of the dye was
much brighter and more durable served to extend the useful life of this colorant. In
fact, Drebbel's scarlet achieved fame by its use as the colorant for the British
redcoats, the trousers of the Russian imperial guard, and for flags, and in particular
for the Star Spangled Banner [57].

4.4.1 Tyrian Purple

Another dye of animal origin achieved fame for quite a different reason: it was so
labor-intensive to produce, so difficult to achieve even small amounts for dyeing
purposes, that it's very scarcity served it very well indeed: it became the Royal
purple, Tyrian purple, the color only royalty could safely wear and not risk being
condemned to death. We are speaking of course, of indigo's first cousin, 6,6′-
dibromoindigo, and other relatives.

The color was, and still is, prepared from several mollusks including *Murex
brandaris*, *Murex trunculus* and *Purpura haemostoma*, found on the shores of the
Mediterranean and the Atlantic coast as far as the British Isles [31, 32]. The
method of extraction and preparation of the colorant was one of the most complex,
time-consuming, and labor-intensive activities carried out by the ancients. The
reasons are (1) it purportedly takes 10,000 shellfish to produce one gram of the
pure colorant (see Fig. 4.7); (2) the colorant itself is not actually present in the
living animal but its precursor must be excised from the snail's hypobranchial
gland; (3) the colorant must be produced by a complex chemical process involving
enzymatic hydrolysis of the dye precursors and subsequent photochemical
oxidation.

Although some archaeologists attribute discovery of the color to the ancient
Minoans (of Crete and other Aegean Islands) around the eighteenth century BCE,

the earliest direct archaeological evidence places dyeworks on the Lebanese coast near the Phoenician city of Tyre around the thirteenth century BCE [58, 59]. Hence the name Tyrian purple. The colorant is a typical vat dye similar to its blue counterpart, indigo. The vatting process involves dissolution of the solid dye by chemical reduction to its leuco, or near-colorless, form by addition of alkali, often calcined eggshells, then immersion of the textile into the dyebath, removal, and subsequent air-oxidation back to its insoluble oxidized purple form now inhering in the textile fibers [60]. Because of the backbreaking yet meticulous work and the constant stench of millions of rotting snails, this could not have been the most favored detail assigned to ancient slaves. However, the dyeworks continued for many centuries into the Christian era, large-scale production ceasing for good with the fall of Constantinople in 1453 CE.

Since Tyrian purple is a natural dye derived from an animal, one might expect that not every sample of this dye is identical, and indeed, this is the case. Koren [60] has shown that there are enormous variations in the percentages of colorants from species to species of the dye snail, as well as the location at which the snails were collected. For example, *Murex trunculus* collected in Spain contained as many as nine pigments, whereas the same species collected in northern Israel contained only five. The percentages of these pigments contribute to the overall hue of the dye since the hue ranges from blue (indigo) to blue-purple (monobromoindigo) to purple (dibromoindigo). In this study, the snail from Israel contained almost no indigo, whereas the Spanish sample contained as much as 40 % indigo. These distinctions have enormous implications for studies on the long-lost ancient color called *tekhelet*, which the Bible mentions as the color of the ceremonial robes of the priests and the ritual prayer tassels worn by the common Israelite. Traditional interpretations have characterized the color as "sky-blue" and it is this color that inspired the color-design of the Israeli flag. However, Koren's analysis of the first-known physical sample of *tekhelet* found at the 2000-year-old excavation at Masada showed that the bluish-purple color was due to a dye from the *Murex trunculus* snail—certainly sky-blue, but the color of the sky at midnight [61, 62]!

The color purple also plays an important role in the development of biological stains. From ancient times, purple was considered to be the most beautiful, stable and precious of all colors. Today, it is critical to the staining of blood. The journey of the Tyrian purple of the ancients to the "Tyrian" purple of William Henry Perkin and beyond is beautifully documented by Krafts, et al. A major stage on this journey was the 1891 discovery by Ernst Malachowski of a borax-alkalinized methylene blue stain that rendered the nucleus of *Plasmodium falciparum*, the malarial parasite, visible for the first time - in all its purple glory [63].

Fig. 4.8 Navajo Dye Chart (© Ella Myers). Toh-Atin Gallery, Durango, Colorado. Key: row 1: scarlet bugler, afterbath from black dye, Brigham tea, brown onion skin, penstemon, sage brush; row 2: juniper mistletoe, red onion skin, Indian paint brush, rubber plant; row 3: alder bark, Navajo tea, wild black berries, rabbit brush; row 4: snake weed, purple larkspur, wild onion, Gambel oak bark; row 5: sumac, piñon pitch and ocher, yellow sweet clover, blue flowered lupine, globemallow, goldenrod, grey chamiso

4.4.2 Natural Colorants from Around the World

Although some colorants, such as mercury(II) sulfide, have been synthesized from ancient times, most colorants before the advent of modern organic chemistry were taken from either plants or animals. There is a long tradition among artists that plant pigments, which are often soluble in the binding medium, must be rendered insoluble by precipitation of the coloring matter on an inert, finely divided, semi–transparent solid such as calcium sulfate or alumina [31, 32]. A lovely garden at the Cloisters, the medieval branch of the Metropolitan Museum of Art in New York City, has a medieval garden which features some of the plant materials used by medieval artists. Among them are Dyer's Greenweed (*Genista tinctoria*), abundant in nearly all the countries of Europe; Woad or "Pastel" (*Isatis tinctoria*), widely cultivated in many European countries until the seventeenth century; Weld or Wold or Yellow Mignonette (*Reseda luteola*), colorant from prehistoric times mentioned by Virgil, Pliny and Vitruvius; Alkanet or Dyer's Bugloss or Orchanet (*Anchusa officinalis*), a perennial plant used from classical times; and Golden Marguerite or Dyer's Anthemis (*Anthemis tinctoria*), a perennial plant that gives a bright yellow color.

Brunello [38] gives a comprehensive list of natural dyestuffs used before the discovery of synthetic dyes, although the earliest printed book on textile dyeing was Gianventura Rosetti's *Plictho de L'arte Tentori...* of 1548, a work containing natural dye recipes using chemical mordants for the dyeing of silk, wool, and linen [64]. A very strong dye tradition emerged in Ethiopia from the fifteenth century onwards, but it was passed on verbally and was in danger of being lost before the research done by Patricia Tournerie [65]. Native American dye traditions were examined by Joe Ben Wheat [56] and Hayes and Perez [66]; there is also some unpublished and graphic data from the Toh-Atin Gallery, Durango, Colorado,

Table 4.1 Cross-cultural use of natural plant dyestuffs–a partial list

Hue	Wheat [56]	Tournerie [65]	Toh-Atin (Fig. 4.11)	Hayes/Perez [66]	Koren [67]
Red/Pink	Madder (*Rubia tinctorum*); cochineal; brazilwood (*Caesalpina echinata*)	*Rubia discolor rubus* spp. *Carthamus tinctorius*	Scarlet bugler (*Penstemon centranthifolius*) Gambel oak bark (*Quercus gambelii*)	Bark mixtures Careless weed (*Amaranthus palmeri, Amaranthus retroflexus*)	Madder (*Rubia tinctorum*) Archil (*Roccella tinctoria*) Alkanet (*Alkanna tinctoria*)
Orange	Orange II (synthetic, imported)	*Lawsonia inermis* (henna)	Indian paint-brush (*Castilleja* genus)		*Lawsonia alba*
Yellow/Brown	Navajo tea plant (*Thelesperma megapotamicum*); juniper	*Croton macrostachys*	Goldenrod (*Solidago* genus)	*Chamiso* spp., *Juglans nigra*	*Crocus sativus Juglans regia*
Green	Horsetail rush (*Equisetum* sp.); wild chokecherry (*Prunus sp.*)	*Rhamnus staddo* (alum mordant)		Black-eyed susan (*Rudbeckia hirta*); wild gourds (*Cuburbita*)	
Blue	Indigo (*Indigofera anil*)	*Carissa edulis delphinium*		Indigo Sorrel roots (*Rumex acetosa*)	Indigo; lichen (*Roccella* spp.) Turnsole (*Chrozophora tinctoria*)
Violet/Purple	Sunflower (*Helianthus* sp.); purple corn (*Zea mays amylacea*)	*Musa sapientum*		Cedar roots (*Juniperus ashei*)	
Black	Logwood (*Hematoxylin campechianum*); complex of sumac, piñon gum, yellow ochre)	*Datura stramonium*	Sumac (Rhus trilobata)	Honey mesquite roots (*Prosopis glandulosa*)	

regarding the use of dyes by the Navajo nation (Fig. 4.8 and Table 4.1). Wheat's book on blanket weaving in the American Southwest is a primary reference work that systematized the vocabulary and developed a consistent chronology of this art form. His meticulous research also reveals the contexts within which technological traditions in the Southwest developed. Koren presented a comprehensive listing of plant dyestuffs used in ancient Israel in a paper presented at the 209th meeting of the American Chemical Society (April, 1995) and subsequently published [67]. His work, the first published critical historical and chemical analysis of the dye-stuffs used in ancient textiles from Israel, opens a window to an understanding of the processes associated with one of the oldest of chemical technologies. The archaeological periods included in the study span about a thousand years from the Hellenistic period to the seventh century CE. An examination of Table 4.1 indicates that there is very little overlap among these traditions, largely because the use of dyestuffs in local and cottage industries was, with some few exceptions, based on what materials were available locally.

4.4.3 Emergence of the Dye Industry

Large-scale use of major natural dyestuffs was made possible by widespread cultivation of the pertinent plant species such as the madder and indigo plants for the dyestuffs themselves, and cactus plants as hosts for the scale insect, *Dactylopius coccus*, that fed upon them. As the appetite for these natural products grew, European entrepreneurs attempted to either improve upon the natural material (e.g., Drebbel's scarlet and Turkey red), or to synthesize dyes in the laboratory.

Two color "cultures" existed side-by-side at the time: (1) the dyehouses, which were generally small and chemically oriented, contributing to the supply side of the colorant industry and discovering new materials, and (2) the printing houses, which generally relied on the designer and the colorist. Colorists, for example the legendary Heinrich Caro [68], were not academically trained, but first served as apprentices, learning by on-the-job training and moving from place to place. They paid close attention to every dye batch and knew how to add just a tiny bit more of the appropriate colorant to bring about the desired effect. For this reason, they were better paid than even top university professors, sometimes by as much as a factor of two [69].

When William Henry Perkin succeeded in synthesizing the dye he eventually called "mauve" in 1856, he was not the first to successfully synthesize a dye—he stood in a long line of other successful chemists going back to at least the beginning of the fourteenth century. If we define a synthetic dyestuff as a colorant which does not occur in nature, and has to be deliberately made by a chemical reaction, then we must include among them substances with as yet undetermined chemical formulas. Farrar [70] has enumerated some of the more notable synthetic dyestuffs as given in Table 4.2.

Some interesting observations on Table 4.2 are pertinent.

Table 4.2 Notable synthetic dyestuffs (1300 – 1856)

Dye	Formula	Discoverer	Date/CI number	Source/synthesis
Orchil (archil); *Pourpre française*	Complex mix of phenoxazones	Florentine trading family in the Levant	1300 CI natural red 28	Lichens (*Roccella tinctoria* and other species) + prolonged exposure to air and ammonia
Saxe blue	Indigo 5,5′ disulfonic acid (also tri and tetra)	Counsellor Barth at Grossenhayn, Saxony	1740 CI 73015 CI acid blue 74 CI food blue 1 CI pigment blue 63(Al salt)	Indigo + H_2SO_4
Red sulfate of indigo	Indigo monosulfonic acid	Walter Crum (1796–1867)	1823	Used in Mulhouse and Yorkshire, has shades similar to logwood
p-Rosolic acid (aurin)	Phenol + oxalic acid = 4-[bis(p-hydroxyphenyl)-methylene]-2,5-cyclohexadien-1-one	FF Runge (1794–1867);	1834 CI 43800	Coal tar
Pittacal	Hexamethylaurine [71]	Carl von Reichenbach (1788–1869)	1834 CI 43875	Isolated from beechwood tar
Rufigallic acid	1,2,3,5,6,7-hexa-hydroxyanthra-quinone	PJ Robiquet (1780–1840)	1835 CI 58600	Gallic acid + H_2SO_4
Picric acid	2,4,6-trinitrophenol	Peter Woulfe JJ Welter M. Guinon (Guinon, Marnas and Bonnet) of Lyons	1771 1799 1845 CI 10305	Nitration of crude phenol
Chrysammic acid	2,4,5,7-Tetranitrochrysazin	Edward Schunck (1820–1903)	1848	Nitration of aloes

(continued)

Table 4.2 (continued)

Dye	Formula	Date/CI number	Discoverer	Source/synthesis
Murexide	Ammonium purpurate; 5,5′-nitrilodibarbituric acid monoammon-ium salt	1776 1818 (1853-1865) CI 56085	Scheele (1742–1786) W Prout (1785–1850)	Uric acid treated with nitric acid and ammonia
Mauve	Mixture of four related aromatic compounds related to the safranines	1856 CI 50245	WH Perkin (1838–1907)	Impure aniline disulfate with potassium dichromate
Magenta (Fuchsine)	Rosaniline hydrochloride	1856 1858 CI 42510 CI Basic Violet 14	AW von Hofmann (1818–1892) J. Natanson, François Emmanuel Verguin	Impure aniline with stannic chloride

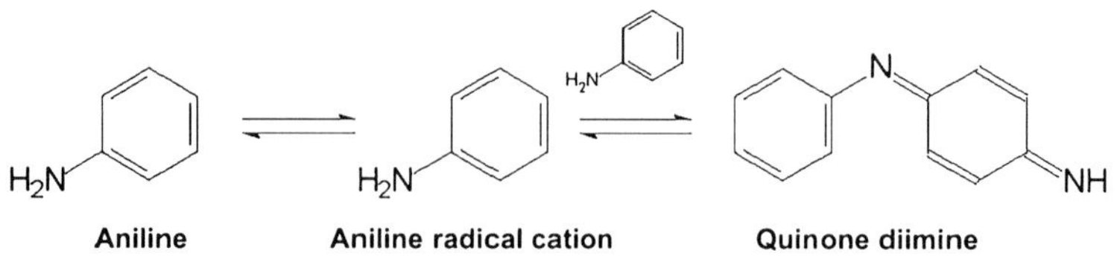

Fig. 4.9 Putative structures of the initial intermediate compounds formed by Perkin in his mauveine synthesis. The dimer quinone diimine reacts with additional aniline molecules to form reactive trimers and tetramers by several pathways, some of which lead to one or more of the seven known mauveine structures (Fig. 4.10)

(1) W. Prout got his uric acid from the excrement of a boa constrictor, one of the sights of London at the time, so that any industrial development had to wait for a more abundant source. Guano imported as fertilizer from the Chincha Islands off the coast of Peru provided that source. However as with other dyes that had natural substances as their starting materials, it would only be a matter of time until the source could not keep up with the demand.

(2) Murexide was a very important color—it did not last long but its volume use during its lifetime was extremely large; it was also the first dye used on a large scale [69].

(3) Of the four coal tar dyes on the market before 1860 (picric acid, mauve, magenta, and rosolic acid), only magenta had a long and profitable life. All four were the parents of other and better colors made from them by further chemical reactions. In this respect the costly mauve was at a grave disadvantage in an increasingly price-conscious market, interested less in silk than in workaday cotton and wool.

Though William Henry Perkin's (1838–1907) synthesis of mauve is "sung" in just about every organic chemistry textbook, there are aspects to the discovery that may not be so obvious. The context of the discovery was actually the profit motive. Perkin was enrolled at the Royal College of Chemistry, an institution backed by industrialists and agriculturalists, under the tutelage of August Wilhelm von Hofmann (1818–1892). So it is no surprise that Perkin abandoned his studies at the age of eighteen to go into business when he realized the value of his discovery. And perhaps Perkin's head for business was even more responsible for his success than his chemical knowledge: at the time, there was no set theory nor was there any knowledge of structural formulas. Perkin was literally working in the dark with molecular formulas that led to essentially trial-and-error reactions. That he succeeded at all is no small miracle—in developing a cheaper pathway to the basic raw material, aniline; in developing ways to make the new dye adhere to cotton as well as silk; in setting a price that would bring him a profit; in staving off competition from a number of quarters, and not least, in surviving preparation of large amounts of his dye in primitive conditions using reagents like fuming nitric acid and hydrogen sulfide [72].

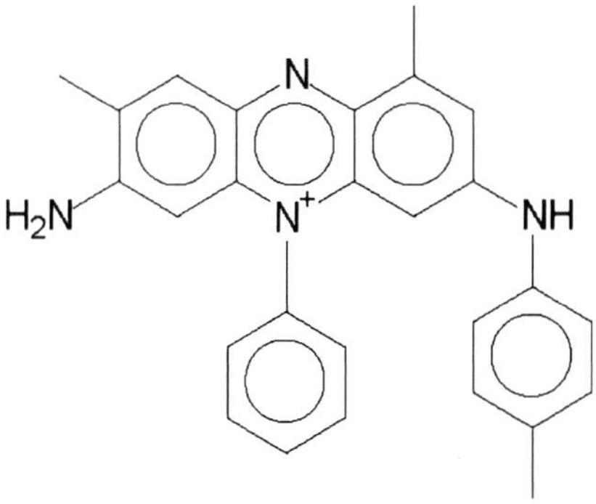

Fig. 4.10 One of the seven structures of mauveine

Fig. 4.11 William Henry Perkin. Williams Haynes Portrait Collection, Chemical Heritage Foundation Collections

Though early on his product was called "Tyrian" purple or aniline purple, the name was changed to mauve a year into production. Perkin's great good luck was twofold: there was great demand for this stunning color in the fashion world, and it was adopted on a large scale by one of the largest calico printers in the British Isles. This effectively quashed two competitors that derived their product from natural sources. The first was Roman purple, sometimes called murexide, that gave it the aura of having come directly from the *Murex* snail, whereas its inglorious source was bat guano imported from Peru (as noted above). The second was French purple (*pourpre Française*, in Table 4.2), made by the action of ammonia and air on various species of lichens. Buoyed by the continuation of mauve mania in the fashion world and his firm's monopoly in Britain, along with his ingenuity in overcoming problems of manufacture, Perkin, by the age of thirty-six, became a wealthy man despite the fact that mauve, in a sense, was a flash in the pan, with production ceasing in 1873, only 17 years after it came on the scene—a victim of its own success in that it encouraged so many competitors trying to find better, less expensive dyes [73]. However, it was not just this one dye that changed the world—it was the train of events that followed it, especially the international competition it engendered. And it did not fall into Perkin's lap. Essentially, Perkin's "decision to get out on the road and solve problems of application, using the methods of both the laboratory and the colorist, and the sweaty laborious work on production processes for intermediates, aided by [his] earlier interest in model-building and mechanics, were all critical components that enabled the aniline dye to replace its lichen-derived competitor" [72].

Perkin began his research largely in the dark. Though he could calculate the molecular formulas of the substances he was working with, it would be another decade before even the faint stirrings of the ideas of Kekulé, Couper, Butlerov and others would lead to a realization of the importance of structural formulas in the understanding of organic chemical reactivity and reactions. Hence, when he began to try to synthesize quinine from N-allyltoluidine by, as it were, brute force—treatment with strong acids and powerful oxidizing agents—he was just plain lucky that these starting materials rearranged themselves into something useful, and he was very astute in recognizing that fact. One of his contemporaries, Louis Pasteur (1822–1895), is famous for his observation that "chance favors the prepared mind".

He was plain lucky perhaps in a more substantial way in that his mentor, Hofmann, had actually discovered the aniline raw material in 1843, knew more about it than anyone else, and had a fertile research mind that suggested a problem that the precocious Perkin could dig into: do humankind a favor by finding a synthetic pathway to quinine, a much sought-after drug for the relief of malaria and some parasitic ailments.

As Perkin continued to try to improve his synthesis after his great discovery, he found that mixing starting materials of aniline, o-toluidine and p-toluidine gave the best results. Early on, he had used only aniline as a starting material, and had obtained a product, but in low yield. It later turned out that his "aniline", as that of almost everyone else, was quite impure, containing a great deal of toluidine's isomers as well, which was a serendipitous accident permitting his discovery. Addition of sulfuric acid and the strong oxidizing agent, potassium dichromate, was a necessary step, and although Perkin did not know it at the time, by this step he was forming the highly reactive aniline cation radical which attacked an unchanged aniline molecule, giving rise to a variety of products [74]. The scheme (Fig. 4.9) shows the initial reaction. The formulas are modern renditions that were unknown in Perkin's time. Oxidative coupling of the quinone diimine with aniline and several of its derivatives gives rise to the corresponding mauveine derivatives via the intermediate formation of the dihydrophenazine derivatives.

Strangely enough, it was not until 1994 that the structure above was finally determined to be one of the correct structures, each differing in the number and placement of the methyl groups bonded to the phenyl groups [75]. Figure 4.11 is a photograph of Perkin in his more mature years.

As Farrar almost poetically remarks [70]:

> *Perkin's mauve was only one of the more precocious of a number of flowers which burst into bloom on the same fine day; and it did not prove to be the longest lived or the most brilliant. Perhaps Perkin's greatest invention was not mauve at all, but technical service. The French colour-makers were dyers themselves; they used their own products and solved their own problems. Perkin was an industrial chemist who had to sell to dyers. If they had problems, he had to help solve them, or they would go back to the natural dyes they knew so well. This was quite a new concept in the chemical industry.*

Revolutionary, as a matter of fact!

In the end, the three most important dyes to come out of Perkin's work were mauve, Britannia Violet, and Perkin's Green. The company developed lake-making procedures, thus enabling the company to introduce aniline dyes to printers of wallpapers and to makers of lithographic and other printing inks. Lakes from mauve and Britannia Violet were used in the production of the British Postage and Revenue Stamps made by the De La Rue Company, who were the first printers to make use of inks containing synthetic coloring matters [76].

As we will see in the next chapter, Perkin's mauve drove three world powers–Britain, France, and Germany–to seek domination in the business of producing colors. Within five years of mauve's appearance, there were already 28 dye manufacturers, not only in the "big three," but also in Austria and Switzerland–many of them destined to become the industrial giants that we know today by such familiar acronyms as AGFA and BASF [77].

References

1. Brown B (1943) Woad—the ancient universal dye. Text Color 65:427–432
2. Hamilton R (2007) Ancient Egypt: the kingdom of the Pharaohs. Parragon, New York, p 62
3. Schweppe H (1997) Indigo and Woad. In: FitzHugh EW (ed) Artists' pigments: a handbook of their history and characteristics, vol 3, National Gallery of ArtWashington, DC, pp 80–107
4. Dweek AC (2002) Natural ingredients for colouring and styling. Int J Cosmetic Sci 24:287–302
5. O'Neil MJ (ed) (2006) The Merck index, 14th edn. Merck & Company, Whitehouse Station, p 5393
6. Marcoux D, Couture-Trudel P, Riboulet-Delmas G, Sasseville D (2002) Sensitization to para-phenylenediamine from a streetside temporary tattoo. Ped Derm 19(6):498–502
7. Lambert JB (1997) Traces of the past: unraveling the secrets of archaeology through chemistry. Addison Wesley, Reading, pp 79
8. Lambert JB (1997) Traces of the past: unraveling the secrets of archaeology through chemistry. Addison Wesley, Reading, pp 59–60
9. Gunn F (1973) The artificial face: a history of cosmetics. David & Charles, Newton Abbot
10. RdeF Lamb (1936) American chamber of horrors. Farrar & Rinehart, New York
11. Kay G (2005) Dying to be beautiful: the fight for safe cosmetics. Ohio State University Press, Columbus

12. Riordan T (2004) Inventing beauty: a history of the innovations that have made us beautiful. Broadway Books, New York
13. Denio A (1980) Chemistry for potters. J Chem Educ 57:272–275
14. Hunt LB (1976) The true story of purple of Cassius. Gold Bull 9:134–139
15. Bell BT (1978) The development of colorants for ceramics. Rev Prog Color Relat Top 9:48–57
16. Brown S (1994) Stained glass: an illustrated history. Bracken Books, London
17. Theophilus (1961) De diversis artibus, manuscript of ca. 1123. Latin text and English translation by CR Dodwell. Thomas Nelson & Sons, London
18. Vogel N, Achilles R (2007) The preservation and repair of stained and leaded glass. In: national park service, 33 Preservation Briefs. Last retrieved on 11 December 2011 from World Wide Web: http://www.nps.gov/history/hps/tps/briefs/brief33.htm, p 7
19. Last retrieved on 11 December 2011 from World Wide Web: http://www.emilfrei.com
20. Pliny the Elder (1855) The natural history. Bostock J, Riley HT (trans.), Crane GR (ed) Taylor and Francis, London
21. Caley ER (1926) The Leyden papyrus X—an English translation with brief notes. J Chem Educ 3:1149–1166
22. Caley ER (1927) The Stockholm papyrus—an English translation with brief notes. J Chem Educ 4:979–1002
23. Hedfors H (1932) Compositiones ad tingenda musiva: [Codex Lucensis 490], herausgeben übersetzt und philologisch erklärt. Almqvist & Wiksells, Uppsala
24. Smith CS, Hawthorne JG (1974) Mappae clavicula, a little key to the world of medieval techniques. Trans Am Phil Soc, New Series, 64:Part 4
25. Richards JC (ed) (1940) A new manuscript of Heraclius. Speculum 15:255–271
26. Thompson DV Jr, trans. (1933) The craftsman's handbook 'Il Libro dell' Arte' by Cennino d'A. Cennini, New Haven: Yale University Press; p 36 (reprint of the English translation: Dover Publications, New York, 1960)
27. Merrifield MP (1966) The art of fresco painting, as practised by the old Italian and Spanish masters, with a preliminary inquiry into the nature of the colours used in fresco painting, with observations and notes. Alec Tiranti, London. 1st edition (1846). This edition (1952; reprinted 1966). Mary Philadelphia (née Watkins) Merrifield (ca. 1804–1889) was a Victorian woman who would have been quite at home in the Renaissance era. It was her publication of a translation of Cennino Cennini's A Treatise on painting which led to her being employed by the British government on researches connected with the work of the Royal Commission on the Fine Arts. The materials accumulated, translated, and embellished with aids for the reader, are cited here and in reference [28]. This very thorough collection of early technical information and recipes on all the arts has still not been superseded. Her other works include handbooks on portrait painting and model drawing, as well as natural history guides. In her later years she learned Swedish and Danish in order to read scientific literature in those languages
28. Eastlake CL (1960) Methods and materials of painting of the great schools and masters. Two volumes. Dover Publications Reprint, New York. Sir Charles Lock Eastlake (1793–1865), not to be confused with his nephew Charles Locke Eastlake, was an English painter and art scholar who made major contributions to the literature of the fine arts through his own writings and translations of works of art of historical value. He was the first Director of the National Gallery, London
29. Merrifield MP, trans. (1967) Original treatises on the arts of painting, vol 2. Dover Publications Reprint, New York
30. Thompson DV Jr (1956) The materials and techniques of medieval painting. Dover Publications Reprint, New York
31. Gettens RJ, Stout GL (1966) Painting materials: a short encyclopedia. Dover Publications Reprint, New York, p 124
32. Gettens RJ, Stout GL (1966) Painting materials: a short encyclopedia. Dover Publications Reprint, New York, pp 163–165

33. Harley RD (1970) Artists' pigments c. 1600–1835. Butterworths, London
34. Feller RL (ed) (1986) Artists' pigments: a handbook of their history and characteristics, vol 1. National Gallery of Art, Washington
35. Roy A (ed) (1993) Artists' pigments: a handbook of their history and characteristics, vol 2. National Gallery of Art, Washington
36. FitzHugh EW (ed) (1997) Artists' pigments: a handbook of their history and characteristics, vol 3. National Gallery of Art, Washington
37. Brunello F (1973) The art of dyeing in the history of mankind. Hickey B (trans.) Phoenix Dye Works, Cleveland, pp xv
38. Brunello F (1973) The art of dyeing in the history of mankind. Hickey B (trans.) Phoenix Dye Works, Cleveland, pp 325–394
39. Orna MV, Mathews TF (1988) Uncovering the secrets of medieval artists. Analytical Chem 60:47A–50A
40. Orna MV, Mathews TF (1981) Pigment analysis of the Glajor gospel book of UCLA. Stud Conserv 26:57–72
41. Cabelli DE, Orna MV, Mathews TF (1984) Analysis of medieval pigments from Cilician Armenia. In: Lambert JB (ed) Archaeological chemistry—III. American Chemical Society, Washington, pp 243–254
42. Orna MV, Lang PL, Katon JE, Mathews TF, Nelson RS (1989) Applications of infrared microspectroscopy to art historical questions about medieval manuscripts. In: Allen RO (ed) Archaeological chemistry IV. American Chemical Society, Washington, pp 265–288
43. Merian SL, Mathews TF, Orna MV (1994) The making of an Armenian manuscript. In: Mathews TF, Wieck RS (eds) Treasures in heaven: Armenian illuminated manuscripts. The Pierpont Morgan Library, New York and the Princeton University Press, pp 124–142
44. Kraft A (2008) On the discovery and history of Prussian blue. Bull Hist Chem 33(2):61–67
45. Kraft A (2011) "Notitia Coeruleo Berolinensis nuper inventi" On the 300th anniversary of the first publication on Prussian blue. Bull Hist Chem 36(1):3–9
46. Mitchell MM, Barabe JG, Quandt AB (2010) Chicago's Archaic Mark (ms 2427) II: microscopic, chemical and codicological analyses confirm modern production. Novum Testamentum 52:101–133
47. Clark RJH (1984) The chemistry and spectroscopy of mixed-valence complexes. Chem Soc Rev 13:219–244
48. Sistino JA (1973) Ferriferrocyanide pigments: iron blue. In: Patton TC (ed) The pigment handbook, vol I, Wiley-Interscience, New York, pp 401–407
49. Day P (1970) Mixed-valence compounds. Endeavour 29:45–49
50. Berrie B (1997) Prussian blue. In: FitzHugh EW (ed) Artists' pigments: a handbook of their history and characteristics, National Gallery of Art, vol 3, Washington, pp 191–217
51. Bader A (2008) Chemistry & art: further adventures of a chemist collector, Weidenfeld & Nicolson, London, pp 123–142
52. Cotton FA, Harmon JB, Hedges RM (1976) Calculation of the ground state electronic structures and electronic spectra of di- and trisulfide radical anions by the scattered wave-SCF-X_α method. J Am Chem Soc 98:1417–1424
53. Plesters J (1993) Ultramarine blue, natural and artificial. In: Roy A (ed) Artists' pigments: a handbook of their history and characteristics, National Gallery of Art, vol 2, Washington, pp 37–65
54. Carlo-Stella C (2006) Mosaics at Saint Peter's Basilica: an eloquent vehicle to express the continuity of sacredness. In: Riehl C (ed) An evolution of the human spirit as seen through mosaic art. Editions du Signe, Strasbourg
55. Rozenberg S (2008) Hasmonean and Herodian palaces at Jericho: final reports of the 1973–1987 excavations. Volume IV: the decoration of Herod's third palace at Jericho. Israel Exploration Society and Institute of Archaeology, The Hebrew University of Jerusalem, Jerusalem, pp 250–274
56. Wheat JB (2003) Blanket weaving in the southwest. University of Arizona Press, Tucson
57. Greenfield AB (2005) A perfect red. HarperCollins, New York, p 141

58. Haubrichs R (2006) Natural history and iconography of purple shells. In: Meijer L (eds) Indirubin, the red shade of indigo. Life in Progress Editions, Roscoff, Ch. 6, p 55
59. Cardon D (2007) Natural dyes—sources, tradition, technology and science. Archetype Publications, London 2011
60. Koren ZC (2008) Archaeo-chemical analysis of royal purple on a Darius I stone jar. Microchim Acta 162:381–392
61. Koren ZC (2011) Tekhelet: announcing the discovery of the first authentic biblical-blue tekhelet from ancient Israel after a millennium and a half of disappearance. Presentation at the International Edelstein Color Symposium, Shenkar College of Engineering and Design. Ramat-Gan, Israel, 27–28
62. Kraft D (2011) Rediscovered, ancient color is reclaiming Israeli interest. The New York Times International, 28 Feb 2011, p A7
63. Krafts KP, Hempelmann E, Oleksyn BJ (2011) The color purple: from royalty to laboratory, with apologies to Malachowski. Biotech Histochem 86:7–35
64. Burns DT, Piccardi G, Sabbatini L (2008) Some people and places important in the history of analytical chemistry in Italy. Microchim Acta 160:57–87
65. Tournerie PI (2010) Colour and dye recipes of Ethiopia, 2nd edn. New Cross Books, London
66. Hayes J, Perez P (1997) Project inclusion: native American plant dyes. Chemical Heritage 15(1):38–40
67. Koren ZC (1996) Historico-chemical analysis of plant dyestuffs used in textiles from ancient Israel. In: Orna MV (ed) Archaeological chemistry: organic, inorganic and biochemical analysis. American Chemical Society, Washington, pp 269–310
68. Reinhardt C, Travis AS (2000) Heinrich Caro and the creation of the modern chemical industry. Kluwer, Dordrecht
69. Homburg E (1983) The influence of demand on the emergence of the dye industry. The roles of chemists and colourists. J Soc Dyers Colour 99:325–333
70. Farrar WV (1974) Synthetic dyes before 1860. Endeavour 33(149–155):155
71. Kaufmann GB (1977) Pittacal—the first synthetic dyestuff. J Chem Educ 54:753
72. Travis AS (1993) The rainbow makers: the origins of the synthetic dyestuff industry in western Europe. Lehigh University Press, Bethlehem 64
73. Garfield S (2001) Mauve: how one man invented a color that changed the world. WW Norton, New York, p 147
74. Heichert C, Hartmann H (2009) On the formation of mauvein: mechanistic considerations and preparative results. Z Naturforsch 64b:747–755
75. Meth-Cohn O, Smith M (1994) What did W. H. Perkin actually make when he oxidised aniline to obtain mauveine? J Chem Soc Perkin Trans 1:5–7
76. Fox MR (1987) Dye-makers of Great Britain 1856–1976. Imperial Chemical Industries, Manchester, p 97
77. McGrayne SB (2001) Prometheans in the lab: chemistry and the making of the modern world. McGraw-Hill, New York, p 22

Chapter 5
Beyond Perkin

As one might expect, following Perkin's discovery, many industrial dyers and chemists began systematic searches for additional aniline colors by mixing aniline with just about anything they could think of. Among these experimenters was François Emmanuel Verguin (1814–1864) who, as early as 1858, discovered a red dye, later called aniline red, made by treating aniline with anhydrous stannous chloride. Others tried to improve on this process, the most successful of whom was Edward Chambers Nicholson (1827–1890), one of A. W. Hofmann's first students and a prominent supplier of the starting material, aniline, through his firm of Simpson, Maule and Nicholson.

5.1 Early Attempts at Dye Synthesis in the Years Following Mauve

Some time in the year 1859 or 1860, Nicholson noticed that when aniline was treated with a strong solution of arsenic acid, the aniline red of Verguin could be obtained in much higher yield and he marketed it under the trade name roseine, although it was commonly called magenta in Britain, and fuchsine in France. Things moved rapidly at this point, perhaps aided by a little industrial espionage. Jules Théophile Pelouze (1807–1867), an expert in benzene manufacture and nitration, put two of his assistants to work to perfect and scale up the arsenic method of producing aniline red. The two, Georges Ernest Camille de Laire and Charles Adam Girard, discovered a new dye through a workman's error, thus adding an exclamation point to Louis Pasteur's (1822–1895) maxim that "chance favors the prepared mind!" Accidentally adding excess aniline to their "aniline red" mixture actually produced a blue product that they patented and named aniline blue.

By this time the center of gravity for color production was already slowly shifting across the Channel to France, particularly to Mulhouse in the Alsace, and

M. V. Orna, *The Chemical History of Color*, SpringerBriefs in History of Chemistry, 79
DOI: 10.1007/978-3-642-32642-4_5, © The Author(s) 2013

to Germany. At first, among the small companies and firms racing to beat one another to the patent office, there was little or no consideration of the scientific aspects of color production. Virtually the only voice crying in the theoretical wilderness was that of A. W. Hofmann, someone who was in on the ground floor regarding Perkin's discovery in 1856. Actually, Hofmann was in an optimum position to carry out systematic scientific research on the new dyes. First of all, he was the world expert on aniline and its reactions; secondly, he had access to pure crystalline aniline red, from which he succeeded in making a series of derivatives for study. He was also closely associated with Pelouze's two assistants, giving him access to aniline blue as well. Using his expertise, he was able to optimize the conditions under which these two priceless dyes could be produced, and this of course, boosted the respectability of organic chemistry tremendously: industrialists could see first-hand how useful scientific knowledge could be. Perkin himself was a beneficiary of Hofmann's research since through it he found out that a mix of aniline with toluidines was necessary for the production of his mauveine.

5.2 Hofmann's Early Theoretical Work

Hofmann began his important work on aniline red by carefully determining its molecular formula through combustion studies. However, since this work was done about 5 years before Kekulé's ideas on structure were published, there was no theoretical basis for going beyond the molecular formula, which provided only knowledge of molecular weight and the ratios of atoms to one another. Missing was knowledge of how these atoms related to one another in a structural way, although one is tempted to hypothesize that Hofmann realized the importance of structure at a very early date. He even began constructing rough models using croquet balls as early as 1865 [1, 2], the year that Kekulé introduced his benzene ring model. In one of his experiments, he actually came very close to structure. By precise analytical measurements, he was able to determine that three phenyl groups (C_6H_5-) added to aniline red to make aniline blue, so if the formula for the hydrochloride salt of aniline red were $C_{20}H_{19}N_3HCl$, then the formula for aniline blue must be $C_{20}H_{19}N_3(C_6H_5)_3HCl$. Where indeed in the molecule the three phenyl groups were located would remain a mystery for quite some time to come, but they were at least three precise pieces in the puzzle. Additional evidence led to Hofmann's interest in the relationship between color and constitution, so that in 1863 he was able to speculate presciently [3]:

> whether chemistry may not ultimately teach us systematically to build up colouring molecules, the particular tint of which we may predict with the same certainty with which we at present anticipate the boiling-point and other physical properties of the compounds of our theoretical conceptions?

Shown below in Fig. 5.1 is what we now know about how aniline red was made from the three aromatic compounds, aniline, o-toluidine, and p-toluidine. The

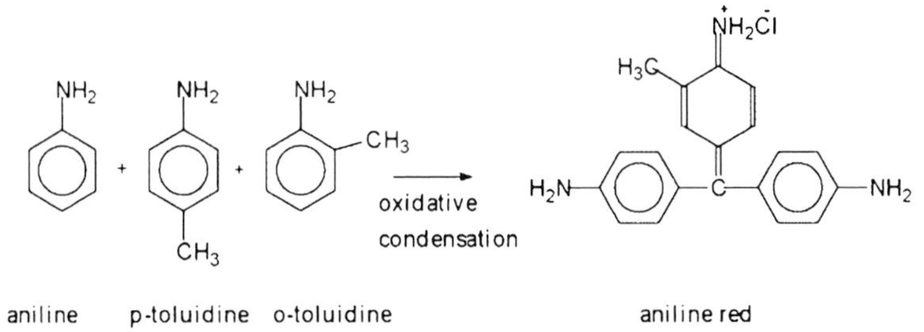

Fig. 5.1 Scheme for synthesis of aniline red from the three starting materials, aniline, p-toluidine and o-toluidine

Fig. 5.2 August Wilhelm von Hofmann. Williams Haynes Portrait Collection, Chemical Heritage Foundation collections

PROFESSOR A.W.HOFMANN (F₀.D.
PROFESSOR IN THE ROYAL COLLEGE OF CHEMISTRY LONDON.

aniline red product is shown as a chloride salt which gives it the properties of a dyestuff [4, p. 87].

5.3 Evolution: From Craft to Science

The decade following Perkin's discovery, which Travis [4, p. 157] aptly calls the magenta era, saw the gradual evolution of the chemistry of color from a craft-based activity to a science-based activity, largely due to Hofmann even though he had little theoretical knowledge at the time. What Hofmann did have was very pure crystals of aniline red which contributed to his solution to the formula problem. He also had a cadre of former students and associates working in industrial environments developing the research results that he pioneered. Figure 5.2 is a photograph of Hofmann in his prime.

Table 5.1 Hydrocarbons from coal tar and development of new dyes

Hydrocarbon	Formula(s)	Period of importance	Key product
Benzene (from light oil distillation)		1859–1868	Aniline blue
Naphthalene (from tar acid distillation)		1870s	Synthetic indigo Azo compounds
Anthracene (from coal tar pitch)		After 1870s	Alizarin

This decade also witnessed the shift of the center of gravity of the dye and color industry to Germany, for a number of reasons that had to do with British and French approaches to business—among them, approaches that limited a firm's ability to engage in long-range planning—leaving a wide open field to the Germans who were not fettered by such limitations. Furthermore, the industrial approach to research was more or less one of stimulus–response, the stimulus being a patent suit, and the response being the research that would stand up in a court of law. Such a response would hardly produce the systematic, purpose-driven research programs and expert process technology that were to grow out of German industry immediately prior to the First World War.

As the decades succeeded one another in the post-Perkin era, certain hydrocarbons became more available and in higher purity, thus driving the search for new colorants via commercially feasible pathways. Not only did chemists become more adept at synthesis and purification, but as we shall see in the next section, they also developed a formidable theoretical base. Table 5.1 summarizes the developing availability of hydrocarbons derived from coal tar and their importance in the synthesis of new dyes.

5.4 Structural Developments

In 1865, at precisely the time that Hofmann was constructing his croquet ball models, Friedrich August Kekulé von Stradonitz (1829–1896) was dreaming. At least, the tale that is told and the myth that has come down to us is that in trying to wrestle with the problem of benzene's puzzling molecular formula, C_6H_6, and reconcile it with a tetravalent carbon atom, he dozed off! Later on, he would write: "I turned my chair to the fire and dozed...the atoms were gamboling before my eyes...all twining and twisting in snake-like motion. But look! What was that! One of the snakes had seized hold of its own tail, and the form whirled mockingly before my eyes. As if by a flash of lightning I awoke; and this time also I spent the rest of the night working out the consequences of the hypothesis...Let us learn to dream…then perhaps we shall find the truth…but let us beware of publishing our dreams before they have been put to the proof by the waking understanding" [5].

Kekulé had made the rounds of the hothouses of chemical ideas in Germany, England and France, studying first at Giessen under Justus von Liebig's (1803–1873) influence, later at Paris with Alexandre Dumas (1802–1870), working with John Stenhouse (1809–1880) in London (where he met William Odling (1829–1921)) and still later with Robert Bunsen (1811–1899) at Heidelberg, where he was a docent. In 1858 he moved to Ghent as a professor and it was during his nine-year tenure there that he made his breakthrough regarding the structure of benzene. However, as with anything else, we stand on the shoulders of those who preceded us, or at least, we move together, side by side, in forwarding new ideas. Kekulé's structural ideas owe much to his association with Odling in London and Charles Frédéric Gerhardt (1816–1856) in Paris, and in 1858, he and Archibald Scott Couper (1831–1892), who was working in Charles-Adolphe Wurtz's (1817–1884) laboratory, came to the same conclusion simultaneously: that carbon atoms were tetravalent and that they had the ability to bond with one another. Although this insight was the breakthrough that opened the doors to a fuller understanding of structure for aromatic compounds, it was not without its problems: it could not then explain the non-observance of ortho-isomers of benzene such as 1,2-dibromobenzene or of catechol or of o-xylene. That would have to wait for theories of resonance and pi-electron delocalization ideas.

However, it did not take long for these burgeoning structural ideas to create a quiet revolution within the following decade, from about 1868 to 1880. During this time, the dye industry moved from a semi-empirical approach to a theoretical approach which bore so much fruit that by 1892, a comprehensive publication [6] came out of the University of Basle that gave the major classifications of dyestuffs that had been produced up until that time (including azines, azo-dyes, oxyquinones, triphenylmethane dyes and quinoneimides) as well as a theoretical introduction to the ideas of chromogens and auxochromes and their relation to molecular structure.

5.5 Adolf von Baeyer and the Synthesis of Indigo

Meanwhile, the Kaiser-Wilhelms-Universität (KWU) was inaugurated in Strasbourg in 1872. To it were recruited numerous young, up-and-coming, talented professors with the idea of bringing immediate prestige to the fledgling university. Among them was Adolf von Baeyer (1835–1917), Fig. 5.3, who was the first to synthesize indigo and give it its correct formula, and later to work with Heinrich Caro (1834–1910), first head of research at Badische Anilin- und Soda-Fabrik (BASF) , on anthraquinone-derived dyes [7]. Although Baeyer's tenure at the KWU was for less than three years, it was there that he got his academic start, and it was there that he met and nurtured such later chemical giants like Emil Fischer (1852–1919) [8, 9].

Fig. 5.3 Adolf von Baeyer.
Williams Haynes Portrait
Collection. Chemical
Heritage Foundation
collections

Although the KWU was Baeyer's first professorial position, he was certainly in the thick of chemical developments long before that. Cognizant of the need to discuss and try out new ideas along the model of the Royal Society in England, he was one of the founders of the German Chemical Society (Deutsche Chemische Gesellschaft). He was also deeply involved in the study of indigo, the plant-derived dye of major industrial importance throughout the nineteenth century. Working on the problem of indigo's degradation products between 1860 and 1865, he realized that the structural building block of indigo was indole, C_8H_7N, information that allowed him to report a useful constitutional formula for indigo as early as 1868 [10]. Baeyer's research on indigo, for a variety of reasons including building new laboratories at the KWU and later at Munich, as well as an admonition from Kekulé, experienced an 8-year hiatus. It was only in 1876, due to collaboration with Caro at BASF, that he took up indigo again. Baeyer and Caro succeeded in devising several synthetic pathways for indigo, none of which were commercially viable mainly due to the great expense of the starting materials, principally toluene. Baeyer was able, nonetheless, to deduce the correct structural formula for indigo in 1883 [4, p. 223]. However, the process that Caro scaled up based upon Baeyer's research was a commercial failure [11, p. 187]. The economic difficulty was overcome when Karl Heumann (1851–1894) discovered in 1890 how to produce anthranilic acid from naphthalene, which was plentiful and

Fig. 5.4 Synthesis of indigo

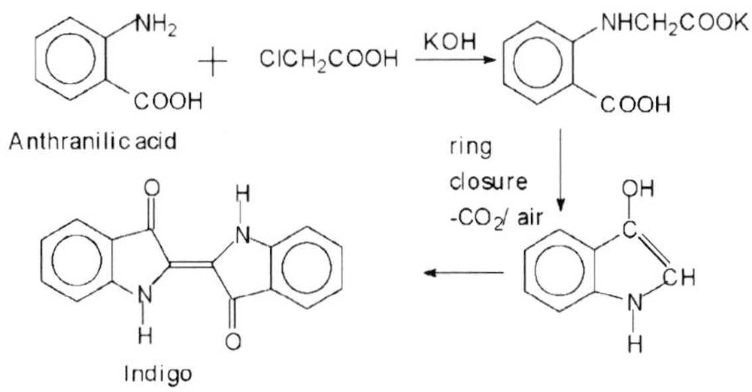

cheap. This led to the first successful industrial process for indigo synthesis in 1897 as shown in Fig. 5.4.

Following this development, the demand for natural indigo withered dramatically, and in fact, by about 90 % within a period of only 15 years. This dealt a lethal blow to Great Britain's Indian monopoly, rendering a lucrative colonial possession into a dependency in a very short period of time. According to Travis [4, p. 226], by 1914 only 150,000 acres of Indian soil were given over to indigo cultivation as compared with about 1,700,000 acres in 1897. The economy in the United States was unaffected by this chemical development because, although indigo had been a viable cash crop in the Carolinas and Louisiana in the eighteenth century, cotton replaced it in the nineteenth century as commercially more successful.

Indigo belongs to that class of dyes commonly known as vat dyes. Generally, dyes of this type are sold as solid materials in the forms of cakes or powders that are insoluble in aqueous media. In order to solubilize the material, it needs to be first reduced to the leuco acid intermediate which is nearly colorless, then to the soluble leuco salt in the presence of an alkali (in ancient times from wood ash, plant ash or eggshells). The textile to be dyed is then immersed in this solution, and upon exposure to air, the dye will re-oxidize to the pigment within the fibers of the textile. The material can be successively dyed to produce a deeper color.

Indigo is one of the few naturally occurring dyes still in use today. One great drawback of most natural dyes has been their notorious lack of fastness, either to light or to washing. Indigo is no exception, but its lack of fastness has actually worked in its favor. What self-respecting teenager would accept a pair of blue jeans that looked pristine and new? It is that worn and faded look, deliberately cultivated, that accounts for indigo's popularity today.

5.6 From Madder to Alizarin: A Convoluted Journey

In the game of cricket a skilled bowler can deliver a ball with a high velocity and elusive spin. For a batsman trying to make contact, the only visual cue for when and where to swing the bat is a split-second flash of the glossy red cricket ball. To

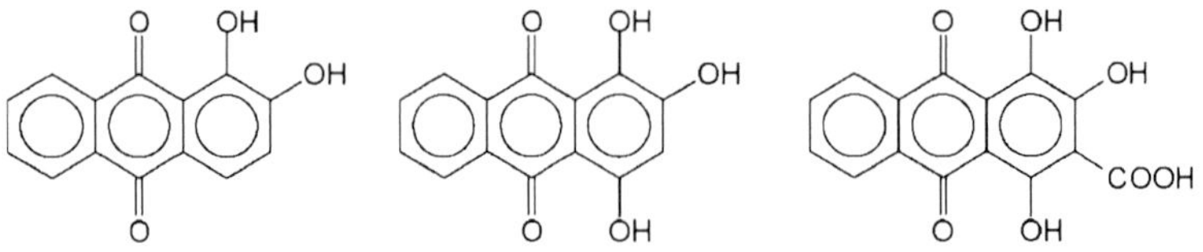

Fig. 5.5 Components of the root of the madder plant, *Rubia tinctorum*

make the ball more visible, early cricket ball makers relied on rose madder to impart a bright red color to the ball's leather cover. Although rose madder is no longer used, the synthetic analogs now favored by cricket ball manufacturers owe their existence to research on this important dye. In 1826, French chemists Jean Jacques Colin and Pierre Jean Robiquet (1780–1840) isolated the primary red coloring agents in rose madder, alizarin and purpurin. Forty-two years later, the German chemists Graebe and Liebermann, prepared alizarin from anthracene and were the first to produce a synthetic substitute for a naturally occurring dye [12].

While indigo was one of the most sought-after of the natural dyes, madder, derived from the ground root of the madder plant, *Rubia tinctorum*, was a close second. Valued from ancient times, it was an important agricultural product throughout the world, principally in India, Turkey, and many parts of Europe, particularly Holland. Naturally occurring madder actually consists of two principal colorant components: alizarin and pseudopurpurin, both belonging to the group of compounds known as anthraquinones after the parent compound, anthraquinone. Their structures are shown in Fig. 5.5.

Alizarin is the major component; purpurin does not occur in the freshly cut root, but only in roots that have been placed in storage, because of the decarboxylation of pseudopurpurin [13]. Although purpurin is a dye in its own right, it is usually considered an undesirable contaminant of alizarin extracted from the madder root [14]. Although almost 30 naturally occurring anthraquinones have been identified [15], alizarin is by far the most commercially viable and therefore the compound for which a viable synthetic route was sought in the 1860s.

Alizarin is a mordant dye, that is, it does not have a strong chemical affinity for textile fibers and therefore requires a fixing agent, or mordant. Commonly metal salts served as these bridging agents between the textile and the dye. Mordanting with a metallic ion was accomplished by adding the mordanting agent to the dyebath itself, by treating the textile substrate with the mordant prior to dyeing, or by treating the already dyed fibers with the mordant. Figure 5.6 illustrates the interaction of the mordant, M, and two bidentate ligands to form a chelate complex.

Here we see a pair of 1-hydroxyanthraquinone molecules forming a neutral 2:1 complex with a divalent metal ion. If the ion, due to electrostatic or ionic bonding interaction is fixed in the fiber, then the dye molecules are also fixed. However, since this is a chemical reaction, then changing the metal ion will change the identity of the complex, and hence, the color. Alizarin, for example, when

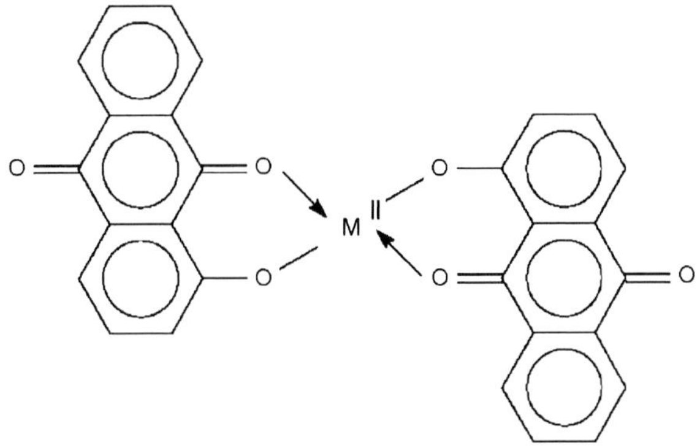

Fig. 5.6 Formation of a 2:1 neutral chelate complex between 1-hydroxyanthraquinone and a divalent metal ion

mordanted with aluminum, tin, chromium or copper will form red, pink, puce brown and yellow brown complexes respectively [16]. Among the colors possible with other mordants are purple, black, and even chocolate. (Please refer back to Sects. 3.5.1 and 3.5.2 for a review of the molecular basis of the color change due to a change in the crystal field splitting energy). Furthermore, the compounds present in the coloring matter or matters possess an unusually stable character, so they can be exposed to the action of various agents for the purpose of improving or modifying the shade [12]. Given its versatility, and the fact that it could impart brilliant colors to cotton and wool, alizarin was a prize worth synthesizing. However, despite much incentive and many studies, it was a very elusive goal, and one that could only be accomplished by a deeper understanding of the structural characteristics of aromatic chemistry. And once again, we find Adolf von Baeyer at the forefront of these efforts, and in fact, long before his great success with indigo.

5.7 Synthesis of Alizarin

After Baeyer had finished his Ph.D. with Bunsen and had worked for a time with Kekulé, he returned to his native Berlin looking for a position, but had little luck. He ended up with a poorly paid lectureship at the Berlin Institute of Technology, also known at the *Gewerbe Institut*. There he found the ideal environment for doing research in natural product chemistry, and his work in this area, though seemingly a dead end, eventually bore fruit through his collaboration with his newly arrived assistant, Carl Graebe (1841–1927) in 1865, the same year as the advent of the benzene ring theory. Graebe began to wrestle with the alizarin problem by studying quinone derivatives for a period of 2 years. It was then that Carl Liebermann (1842–1914), a former student of Baeyer's, arrived, and there, under Baeyer's direction and utilizing his suggestions, he and Graebe found that the parent compound for alizarin was not naphthalene, as they had previously thought, but anthracene, a minor component of pitch. Although anthracene could be readily oxidized to anthraquinone by nitric acid, direct methods for then introducing the two hydroxyl groups were lacking, and the solution they found by bromination turned out to be very expensive. Nevertheless, they were successful with this method in synthesizing a natural dyestuff for the first time, albeit by an inefficient and expensive route that had a very low yield. However, as Travis [4, p. 176] remarks, "a natural product of considerable complexity and of enormous commercial value had been created in the laboratory."

Although successful synthesis of natural dyes of historic interest was viewed as a chemical feat of the first order, no naturally occurring dye could stand up to the stringent requirements eventually set by the dye industry: consistent and controllable hue, lightfastness, and fastness to washing with soaps and other alkali agents, and later on, detergents. Alizarin was no exception. It was recognized early on that the natural material, madder, had some excellent qualities but also some serious drawbacks. On the plus side, madder was more fast than most other natural colorants, it gave strong bright colors, and it was relatively inexpensive because Europeans could grow it themselves. But in an industry that required great precision, the big problem was that the madder roots varied greatly in quality. The dye was also sensitive to both alkalinity and temperature and had a tendency to give an orange shade [17, p. 28]. Although the employment of the synthetic product, alizarin, could obviate some of these problems, others were inherent in the nature of the dyestuff itself. It is no wonder, then, that chemists seeking new and better dyestuffs would first try to characterize and synthesize the natural colorants, but then use their structures as models for developing new colorants with more desirable properties.

Nevertheless, the advent of synthetic alizarin had disastrous consequences for the madder growers and carmine-cochineal cultivators. Almost overnight, the *Dactylopius coccus* growers of the Canary Islands were bankrupted; the madder farmers of Southern France had to convert their fields to potatoes and (perhaps happily for some) grapevines. And most important of all, the center of gravity of the dye industry shifted once and for all from England to Germany, the major patent-holder for the new technology.

5.8 Conclusion: Legacy of the Dye Industry

Two years after the successful synthesis of alizarin in the laboratory, another major breakthrough occurred in Germany, but this time of an organizational, not a technical, nature. In 1867, presided over by none other than A. W. von Hofmann, the German Chemical Society (Deutsche chemische Gesellschaft) was founded. This society was to have far-reaching effects on two major fronts: (1) through its journal, *Berichte*, it became a vehicle for the spread of modern chemical ideas and the recording of the enormous amount of experimental work leading to the rapid determination of the structural formulas for thousands of chemical compounds, and (2) through its petition to the government to amend its draft patent law, it fostered the development of in-house industrial research laboratories leading to the rise of an innovative, science-based chemical industry in Germany [18].

Travis [19], in his Dexter Award Address: "What a Wonderful Empire is the Organic Chemistry" delivered before the American Chemical Society's Division of the History of Chemistry, presented a "Timeline for the Synthetic Dye Industry" which marks the key steps in the industry's development in the latter half of the nineteenth century.

Timeline for the Synthetic Dye Industry, 1856–1900.

1856   : William Henry Perkin in London discovers a purple aniline dye, from 1859 known as mauve. Aniline is made in two steps from coal-tar benzene.

1858–60 : A red dye is made from commercial aniline (containing toluidines). The process is developed by French and British chemists. The colorant is known as magenta, fuchsine, etc. and in 1861 is converted into a blue dye, aniline blue.

1863   : By substitutions into amino groups of magenta, A. Wilhelm Hofmann discovers the Hofmann's violets in May 1863. Since structural formulas are not available for aromatic compounds they are represented by constitutional formulas based on simple "types" of groupings of atoms.

1865   : Friedrich August Kekulé announces his benzene ring theory. This makes it possible to draw the structural formulas of simple aromatic chemicals.

1868   : Carl Graebe and Carl Liebermann in Berlin find that the natural product alizarin is an anthraquinone derivative of the aromatic hydrocarbon anthracene, establish the partial structure of alizarin and a route to its synthesis. This represents the first synthesis of a complex natural product in the laboratory.

1869   : Heinrich Caro at BASF and William Perkin independently discover commercial routes to synthetic alizarin. Patents are filed in London during June 1869. Manufacture begins in England and Germany during 1869–1870 and leads to the decline in the cultivation of madder. This lays the foundation of modern science-based industry and industrial-academic collaboration.

1873   : Unable to compete with the German manufacturers of synthetic alizarin, Perkin retires from industry.

1874   : Academic chemist Adolf Baeyer, at Strasbourg, and industrial chemist, Heinrich Caro, at BASF, jointly publish the modern structure of alizarin.

1875   : Introduction of azo dyes that contain the atomic grouping $-N = N-$, based on academic and industrial research.

1877   : Comprehensive patent law introduced in Germany, after consultation with the dye industry. It is the most advanced system in the world for protecting chemical inventions.

1883   : Adolf Baeyer at Munich draws the modern structural formula for indigo.
1889   : Central Research Laboratory, designed by Heinrich Caro, opens at the BASF Ludwigshafen factory. The industrial research laboratory becomes a formal business unit.
1897   : BASF and Hoechst in Germany commence the manufacture of synthetic indigo. This leads to the collapse of the natural indigo trade.
1900   : Germany and Switzerland are the leading dye-making countries and control most of the world market.

By 1900, the dye industry had matured, at least in Germany, in such a way that its influence was to say the least, historic. It did not spring to life unexpectedly and in a vacuum, but it evolved with the help of and to meet the needs of other industries such as heavy chemicals, illuminating gas, and textiles. It also owed its existence to the early pioneers in organic chemical research such as Justus von Liebig (1803–1873) and his students at Giessen. It grew rapidly because of the capitalistic system with intense competition among dye companies, investment of venture capital nineteenth century style, and the freedom to embark on risky ventures with great rewards if they met with success. The industry was embedded in an economic system that, in enticing greater consumption, led to frantic research, which led in turn to an increased pace of organic chemical discovery [20], and to innovations particularly in process [21].

Here are some of the areas that were completely transformed by the dye industry:

- *Fashion.* Prior to aniline dyes, color austerity was the order of the day: only the rich could afford the rainbow of colors available in the natural dyes. But the ordinary person's desire for the riot of color, now inexpensive and available to all, coupled with a clarity, variety, and fastness not found in the natural dyes, greatly stimulated dye research;
- *The educational structure of industrial society.* The needs of the dye industry stimulated educational reforms that were already under way at the university level so that at the end of the nineteenth century, Germany had the finest system of scientific and engineering education worldwide—and this system influenced higher learning in all parts of the world;
- *The social structure.* By employing large numbers of academically trained chemists in its plants and by growing a white-collar managerial proletariat, the dye industry displaced the independent tradesman and raised the educational level of the masses; this carried over into other disciplines as well;
- *Political action.* The dye industry organized and supported lobbies, exerted strong influence on legislation, and helped establish a model patent system;
- *Industrial research.* The dye industry originated and developed the industrial research laboratory and the research team, an organizational structure that prevails to this day;

- *Power*. Through its hegemony in the dye industry, Germany became the foremost industrial power in Europe, leading to its almost becoming the foremost power in Europe. World War I was in actual fact a battle of technologies that promoted growth in other branches of the chemical industry, while the dye industry itself declined in importance [20, pp. 149–151].

Not all of these developments were unequivocally positive. Through the dye industry, Germany became a giant war machine, manufacturing explosives, poison gases, photographic film, drugs, natural product substitutes, and a whole host of other products born of its technology. The industrial juggernaut also led to massive pollution of the air and of water sources; it led to a scarring of the earth; above all, it led to an unprecedented increase in chronic diseases, to deaths by heavy metal poisoning, to birth defects, and so much more. The rivers that ran red with industrial effluent also ran red with blood. We might say that the developments in the European dye industry of the late nineteenth century steered the course of twentieth century history right up to and beyond World War II.

References

1. Hofmann AW (1865) Proc Roy Inst 4:401–430; pp. 416
2. Ihde A (1984) The development of modern chemistry. Dover, New York, p 307
3. Hofmann AW (1863) On aniline-blue. Proc Roy Soc 13:9–14
4. Travis AS (1993) The rainbow makers: the origins of the synthetic dyestuff industry in western Europe. Lehigh University Press, Bethlehem
5. Japp R (1898) Kekulé memorial lecture. J Chem Soc 73:100
6. Nietzki R (1892) Chemistry of the organic dyestuffs. Gurney and Jackson, London
7. Stadler A-M, Harrowfield J (2011) Places and chemistry: Strasbourg—a chemical crucible seen through historical personalities. Chem Soc Rev 40:2061–2108
8. Steinmüller F (1993) 1905 Nobel laureate: Adolf von Baeyer, 1835–1917. In: James L (ed) Nobel laureates in chemistry: 1901–1992. American Chemical Society, Washington and Chemical Heritage Foundation, Philadelphia, pp 30–34
9. Huisgen R (1985) Adolf von Baeyer's scientific achievements. Angew Chem 25:297–311
10. Baeyer A (1868) Über die reduction des indigblaus. Ber Dtsch Chem Ges 1:17–18
11. Reinhardt C, Travis AS (2000) Heinrich Caro and the Creation of Modern Chemical Industry. Kluwer, Dordrecht
12. Schunck E (1860) On the colouring matters of madder. Quart J Chem Soc 12:198–221 (Chemical Abstracts Service Calendar Essay)
13. Hill R, Richter D (1936) Anthraquinone colouring matters: galiosin, rubiadin, primaveroside. J Chem Soc, 1714–1719
14. O'Neil MJ (ed) (2006) The Merck index, 14th edn. Merck & Company, Whitehouse Station
15. Schweppe H, Winter J (1997) Madder and alizarin. In: FitzHugh EW (ed) Artists' pigments: a handbook of their history and characteristics. Oxford University Press, New York, pp 109–142 Vol. 3
16. Peters RH (1975) Textile chemistry, Vol. III: the physical chemistry of dyeing. Elsevier, New York, p 649
17. Greenfield AB (2005) A perfect red. HarperCollins, New York
18. Johnson JA (2008) Germany: discipline–industry–profession. German chemical organizations, 1867–1914. In: Nielsen AK, Štrbáňová S (eds) Creating networks in

chemistry: the founding and early history of chemical societies in Europe. RSC Publishing, Cambridge, pp 113–138

19. Travis AS (2008) What a wonderful empire is the organic chemistry. Bull Hist Chem 33(1):1–11 (reproduced with permission of the author and of the division of the history of chemistry of the american chemical society)

20. Beer JJ (1981) The emergence of the German dye industry. University of Illinois, Urbana

21. Hornix WJ (1992) From process to plant: innovation in the early artificial dye industry. Br J Hist Sci 25(1):65–90

Chapter 6
Major Analytical Techniques Based on Color: Volumetric Analysis; Chromatography; Spectroscopy; Color Measurement

6.1 Volumetric Analysis

An old adage among analytical chemists of a "certain age" is that when you see a color change, you can be sure that it marks a chemical change. While this observation may not be entirely true in every instance, it certainly paid homage to the human eye as the detector *par excellence* in the early days of chemistry. Even the yellowing of newspaper is a visible change associated with the darkening of lignin, one of paper's components, by exposure to light and air [1]. The fading of dyes is another familiar chemical change marked by a color change. There are many others that are part and parcel of daily life.

Among the more familiar chemical changes even today in academic laboratories is the change of an acidic solution from colorless to a pinkish tone when titrated with a base in the presence of the indicator, phenolphthalein. While the first titrations did not rely on color change, but rather on the cessation of effervescence or precipitate formation, that soon changed with the introduction of new techniques by Joseph Louis Gay-Lussac (1778–1850) and Karl Friedrich Mohr (1806–1879). The former measured the efficiency of bleaching powder by observing its decolorizing action on an indigo solution [2]. The latter, who might rightly be called the "father of volumetric analysis," was also the inventor of the Liebig condenser (eponymously popularized), the laboratory cork borer and the burette that bears his name. Mohr cleverly replaced precipitation titrimetry in the analysis of chlorides by titration with silver nitrate by introducing potassium chromate as an internal indicator, relying not only on the color development of silver chromate but also on its greater solubility in aqueous solution [3]. His 1855 textbook, *Lehrbuch der chemisch-analytischen Titrirmethode*, was very influential in popularizing volumetric analysis. In it, he enumerated the properties of some of the better plant pigments to use as indicators [4], introduced the concept of the back-titration, and suggested the use of equivalents rather than mole quantities for analytical purposes. A decade earlier, in 1846, E. Margueritte had proposed the use of permanganate solution to estimate the amount of iron in a sample, and in 1853,

M. V. Orna, *The Chemical History of Color*, SpringerBriefs in History of Chemistry, DOI: 10.1007/978-3-642-32642-4_6, © The Author(s) 2013

Robert Wilhelm Bunsen (1811–1899) had introduced iodometric methods of analysis [5].

A popular analytical chemistry textbook [6] outlines the chemistry of titration by saying up front that multiple equilibria are involved in any titration and these must always be taken into account. Ideally, the reagent-analyte reaction must go to completion, which limits the amount of dilution that can be tolerated in the solution so that the equilibrium not shift back toward unreacted forms. A titration curve, that plots pH versus amount of titrant added, can be then analyzed to determine the end point, i.e., the volume at which all the analyte has reacted and at which the curve has its steepest slope. At this point, a jump in the pH occurs with a very small addition of titrant, and the jump can be detected visually by the addition of a dye that changes colors, a colorimetric indicator. For an indicator to work as a marker of the end point in a neutralization titration, ideally its pK_a should be nearly the same as the end point pH. (A major assumption in this method is that the end point of the titration and the equivalence point, or stoichiometric point, of the chemical reaction are the same. If they are not, or nearly so, then an inappropriate indicator has been chosen and the color change could take place either before or after the end point is reached, introducing a major source of error). To choose an indicator, it is wise to consult a standard handbook that lists available indicators, along with their chemical names, pK_a values, pH ranges, λ_{max} in nm, and the color change to be expected. Normally the types of indicators are divided into indicators for aqueous acid–base titrations, mixed indicators, fluorescent indicators, and oxidation-reduction indicators. Among some commonly used acid–base indicators are phenolphthalein, first prepared by Adolf von Baeyer in 1871 [7], and phenol red by Friedrich August Kekulé in 1872 [8].

Titrations that use the human eye as the primary detector are based on color change. Other types of titrations that rely on color change are oxidation-reduction, precipitation, and complexometric titrations. Other detectors indicate voltage or other types of changes such as potentiometric titrations, conductometric titrations and amperometric titrations—all of which require additional instrumentation—and may be quite colorless.

Novel extensions of the use of acid–base indicators to monitor other changes have appeared in the literature recently. Mendes [9] described a colorful ion-exchange pedagogical experiment in which the use of a resin with an adsorbed acid–base indicator allowed students to follow the progress of the ion-exchange front along a column. In a sophisticated yet colorful undergraduate laboratory experiment, Pantaleão et al. [10] described the use of a hollow fiber membrane contactor (HFMC), a device where mass transfer occurs between two phases without dispersion of one phase within another, with the fluids passing on opposite sides of a membrane, to study CO_2 absorption into water. The absorption is followed visually by the change in color of bromothymol blue indicator using a purpose-built HFMC, and enables students to experimentally estimate CO_2 solubility and compare the data with a reference. Mohr's method has come a long way!

6.2 Chlorophyll and Chromatography

One of the great perks of living in New England, despite some terrible winter weather conditions, is the change in tree leaf colors during the fall that draws thousands of "leaf peepers" from all over the world to witness this annual recurring event. The magnificent reds, oranges, and yellows that gradually appear in deciduous trees in September and October, as we all know, is due to the breakdown of the green chlorophyll that masks these colors so that they can appear in all their glory for just a small window of time. But these colors do not have merely an aesthetic purpose. In one fairly recent study, the reflectance, transmittance and absorption spectra of *Acer platanoides* (Norway maple) leaves were recorded in the progress of full-term autumn senescence and compared with absorbance spectra of extracts from and pigment contents in these leaves. The observed changes in leaf spectra and development of the intense yellow color of the leaves are considered as related to the changes in the light depth penetration in the photosynthetic tissues accompanying chlorophyll breakdown. The study suggested that carotenoids serving as effective light traps (in spite of their relatively low concentrations) are able to provide protection against the harmful effects of blue-light irradiation. It can explain the physiological significance of their retention in leaves up to terminal stages of the senescence process [11]. And not only that, if these fall leaves appear to glow in the process, it is not your imagination. Another study [12] determined that the leaves really are fluorescing due to the protective action of the carotenoids contained therein. By trapping light in aging leaves, carotenoids boost photosynthesis in degraded chlorophyll. They also protect the chlorophyll from harmful UV radiation by dissipating trapped light out of the cell, creating a glow.

The glow of chlorophyll and other green leaf components has suffused the name of Mikhail Tsvet, the hero and inventor of chromatography, down through the years, although he never got credit for his discovery during his lifetime. Tsvet actually utilized the fluorescence phenomenon described above in his own research almost a century before.

As its name implies, chromatography literally means "color writing". The principle of the technique has a long history, but at the same time, this history does not necessarily have anything to do with color. The best-known modern household application is the use of filters to remove unpleasant odors and tastes from drinking water. Typically, the liquid water (mobile phase) is passed through a column of finely divided material such as charcoal (stationary phase). Impurities in the water will have an affinity for the particles of charcoal and will adsorb to their surface.

As early as the late eighteenth century, this method was in use by Carl Wilhelm Scheele (1742–1786) to adsorb gases on charcoal, and by sugar chemists to clarify sugar solutions. In the 19th century, petroleum chemists were using the same technique in a more refined way: they would filter crude petroleum through a column full of fuller's earth or other inert adsorbing material. By removing the liquid that filtered through the entire column at different time intervals, or by removing the material from the column at different levels through little portholes, they were able to prepare different fractions of the petroleum that had different physical and chemical properties [13].

The breakthrough experiment came in 1906, although hardly anyone noticed until decades later—and then, as they say, the rest is history. At that time, the Italian-Russian botanist, Mikhail Tsvet (1872–1919), was trying to separate the plant pigments, chlorophylls and carotenoids, by adsorbing them from a petroleum ether solution onto a solid material. Let us read his own words as he describes his discovery [14]:

> The most suitable adsorptive materials were precipitated calcium carbonate, inulin, or sucrose (powdered). If the petroleum ether chlorophyll solution was then shaken with the adsorptive material, the latter carried down the pigment, and with a certain excess of this, only the carotene remained in solution, escaping adsorption. In this way a green precipitate and a pure yellow, fluorescence-free carotene solution were obtained (test for fluorescence in my luminoscope [Zeitschrift für physikalische Chemie (1901) 36, S. 450; Diese Berichte (1906) 24, S. 234]). This carotene solution showed a spectrum with absorption bands at 492–475 and 460–445 nm. If it was shaken with 80 % alcohol, the lower alcohol-water phase remained completely colorless.

> The green precipitate was then brought onto a filter and carefully washed with petroleum ether to separate the last traces of carotene. The filtered yellow liquid could be immediately regenerated with bone meal. Then the precipitate was treated with petroleum ether containing alcohol, which completely decolorized it and gave a beautiful green solution which could then be separated by 80 per cent alcohol by the method described by Kraus [Kraus K (1875) Flora S. 155]. The petroleum ether phase, colored blue-green, contained chiefly the chlorophyllines, while the lower yellow phase contained chiefly the xanthophylls.

> If the petroleum ether solution of the chlorophyll was treated with the adsorption material not in excess, but in portions until the fluorescence vanished, then along with the carotene the xanthophylls also remained in solution. These could be freed from carotene by again treating the decanted solution with the adsorption material and liberating the pigment from the resulting adsorption compound with petroleum ether containing alcohol. The solution of xanthophyll mixture thus obtained shows the following absorption spectrum: 480–470 and 452–440 nm. If it was shaken with 80 % alcohol, the pigment remained almost completely in the alcoholic phase.

> The physical interpretation of the adsorption phenomena that we have considered will be discussed elsewhere. However, here we can mention some related regularities and the resulting applications. The adsorption material saturated with a pigment can still take up another member of a certain mixture and hold it firmly. Substitution can also occur. For example, the xanthophylls can be partly displaced from their adsorption compounds by the chlorophyllines, but not the reverse. There is a definite adsorption series according to which the substances can substitute. The following important application comes from this rule. If a petroleum ether solution of chlorophyll filters through a column of an adsorption material (I use chiefly calcium carbonate which is firmly pressed into a narrow glass tube), the pigments will separate according to the adsorption series from above downward in differently colored zones, and the more strongly adsorbed pigments will displace the more weakly held ones which will move downward. This separation will be practically complete if, after one passage of the pigment solution it is followed, by a stream of pure

solvent through the adsorbing column. Like the light rays of the spectrum, the different components of a pigment mixture in the calcium carbonate column will be separated regularly from each other, and can be determined qualitatively and also quantitatively. I call such a preparation a chromatogram and the corresponding method the chromatographic method. In the near future I will give a later report on this. It is perhaps not superfluous to mention here that this method in its principle and also in the exceptional ease of carrying it out has nothing to do with so-called capillary analysis.

In these remarkable paragraphs, and in particular the last one, Tsvet described the following for a mixture of several components of green leaf material:

- differential adsorption on a stationary (solid) phase,
- differential solubility in a liquid phase involving more than one solvent,
- displacement of one adsorbed species by another on a solid phase,
- an adsorption series,
- separation of components on a chromatography column,
- elution of the various chromatographic zones,
- qualitative analysis,
- quantitative analysis.

In naming his preparation a *chromatogram*, and in calling his method *chromatography*, he unknowingly laid the foundations for one of the most powerful analytical tools ever devised—now a multimillion dollar industry that has pushed detection and measurement limits down to parts per trillion, part of the chemistry curriculum in every undergraduate and graduate department of chemistry, and consisting of many more chromatographic methods than simply visualizing a separation of colored analytes on a transparent column. One must wonder whether this method would ever have developed the way it did if chlorophyll were colorless?

Tsvet followed up this paper, in which he uses the word "chromatography" for the first time, with a second paper in the same volume of the same journal [15]. 5 years earlier, he had described his method in a Russian journal, and a little later, in a Polish journal, but these publications were largely unavailable in Western Europe.

Tsvet's life was marked by a series of losses—of family, of educational credentials, and eventually even of teaching posts and research career. His Italian mother died soon after he was born, and his father, a Russian diplomat, left him in the hands of a nursemaid in Switzerland when he was recalled to Russia. Tsvet earned his doctorate at the University of Geneva, but when he moved to Russia, his degree was not recognized and he had to earn a master's degree at the University of Kazan. Later he held a number of teaching posts in Poland, during which he obtained a second doctorate at the University of Warsaw in 1910. After only four tranquil research years, his career ended because of war—a series of evacuations from Warsaw to Moscow to Nizhni Novgorod to Estonia and finally to Voronezh, where he died of heart failure at the age of forty-seven. His great scientific strength lay in the fact that he was both a botanist and a chemist who was deeply interested in the molecular structure of plants, a fact that led to his great interest in chlorophyll. Figure 6.1, a 1972 Russian circulated cover, depicts Tsvet in this dual role by the attribution above his name: "Russian scientist: botanist and chemist".

Fig. 6.1 A portrait of Mikhail Semyonovich Tsvet (1872–1919) on a 1972 Russian circulated cover. From the collection of Daniel Rabinovich, with his kind permission

Chromatography has an interesting aftermath in the annals of the Nobel Prize in Chemistry. The first is the failure of the 1915 Nobel Laureate, Richard Martin Willstätter (1872–1942), to use Tsvet's method successfully because of problems with his choice of adsorbing species. However, he and his students went on to develop many of the separation techniques that made the identification of natural products possible. His abiding interest in chlorophyll eventually led to the Nobel Prize: he was able to show that there were only two types of chlorophyll, a and b, and that magnesium was an integral part of the chlorophyll structure [16]. His work with naturally occurring colored compounds did not stop there: he eventually was able to distinguish among and identify orange carotene, yellow xanthophyll, and red lycopene [17].

The great chromatography success story was the development of liquid–liquid chromatography by Archer John Porter Martin (1910–2002) and Richard Laurence Millington Synge (1914–1994), who shared the 1952 Nobel Prize in Chemistry for this achievement. Martin, assisted by Synge, built a liquid–liquid countercurrent extraction apparatus equivalent to 200 separatory funnels in order to separate acetylamino acids—and failed. They decided then to hold one of the liquid phases (water) stationary on an inert solid support and to move the other immiscible liquid (chloroform)—and thus liquid–liquid partition chromatography was born. And it even was colorful: they used methyl orange as an indicator because it forms bright

Fig. 6.2 Starch chromatography—A.J.P. Martin and R.L.M. Synge—Nobel Prize 1952. Royal Institute of Chemistry 1877–1977. Stamp issued by Great Britain in 1977

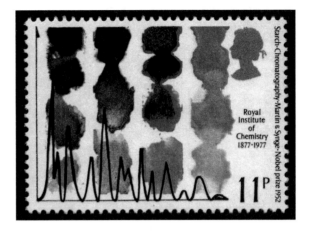

red bands over an orange background to show the separation of the different acetylamino acids [18, 19].

Figure 6.2 is a 1977 British stamp commemorating the 25th anniversary of their Nobel Prize along with the centenary of the founding of the Royal Institute of Chemistry.

It is a curious thing that when they first read their paper to the biochemical society announcing this discovery, they were greeted with a resounding silence and no questions. In Martin's own words, "it raised not a flicker of interest". And some years later, when they had begun the first investigations into substituting a vapor phase for the liquid mobile phase—the beginnings of gas chromatography with a colorful twist as well—they experienced the same reaction. It is fortunate for the future development of these extraordinary techniques that someone in Stockholm was paying attention [20]!

6.3 Spectroscopy

"Spectroscopy" literally means to "look at the spectrum". And as we know well, if we look at something long enough, then our knowledge and understanding of this something gradually becomes transformed. In the words of Vincent van Gogh, "It is looking at things for a long time that ripens you and gives you a deeper understanding." And there are different ways of knowing and understanding. There is factual knowledge: I know that potassium permanganate is purple and discolors in a hydrogen peroxide solution. There is theoretical knowledge: I know that permanganate is purple because of charge transfer transitions and that it is a powerful oxidizing agent that becomes near colorless when Mn(VII) is reduced to Mn(II). There is intimate knowledge: I know that potassium permanganate can stain my fingers brownish-black because I have used it often as a titrant for analyzing iron samples. And there is mystical, or spiritual, knowledge: I am in awe of the gift of permanganate because of its lovely color, because of its action as an antifungal agent, because of its color reactions that give me pleasure in the process of being useful, etc. None of these ways of knowing is sequential—we can experience the mystical sense of something perhaps long before we understand how it works.

And such is the case with spectroscopy. First came the wonder, the marveling at the multicolored arc of the rainbow, first mentioned in the Bible as the sign of God's covenant with humanity (Genesis 9:13). Then came cursory observations until Newton's famous experiment described in Chap. 2, Sect. 2. William Hyde Wollaston (1766–1828) improved on Newton's method by using a narrow slit instead of a round opening for the light, and as a result may have been the first person to observe a dark line spectrum—the solar spectrum interrupted by a series of dark lines parallel to and actually images of the slit [21, 22]. About 15 years later, the German physicist/optician Joseph von Fraunhofer (1787–1826) studied these lines in much greater detail, identifying over 570 of them and mapping them out meticulously, including the measurement of their wavelengths—to seven significant figures. He designated what seemed to be major lines with letters of the alphabet, and gave other symbols to weaker lines [23]. Among the lines, he noticed a very prominent yellow doublet that fell at "D" in his alphabet, so he called them D_1 and D_2. They turned out to be the sodium-D line that we use to this day for many physical measurements and analytical procedures—still designated with his notation. He did not understand the significance of these lines, but he knew what he wanted: he used the dark lines as reference points and measured the refractive index for various glasses with the accuracies needed for achromatic lens design, in the process experimenting with different types of glasses in order to find some with matching dispersion properties. Sparked by a desire to construct better optical instruments he laid a firm foundation for the development of modern spectroscopy [24].

Fraunhofer's life was paradigmatic of poor boy made good—he was the tenth and last son of a poor glass-grinder who could not afford to send him to school. He was orphaned at the age of twelve and was apprenticed without pay to another grinder, this time of lenses. Only when his house collapsed was he brought to the attention of Joseph von Utzschneider (1763–1840), a privy counselor and director of the Benediktbeuren optical institute, who must have noticed talent when he saw it. He furnished Fraunhofer with books on mathematics and optics, and gave him the means to buy himself out of his oppressive apprenticeship. Later he took Fraunhofer into his employ and eventually made him manager and partner due to his excellent work. Though Fraunhofer never had any formal education, he became professor royal when the optical institute moved to Munich, and he eventually received an honorary doctorate in 1822 from the University of Erlangen. His great fame lies in being the initiator of spectrum analysis, the consequence of studying the chromatic diffraction of different glasses. He also accomplished an important theoretical work on diffraction and established its laws; he later made and used diffraction gratings with up to 10,000 parallel lines to the inch, ruled by a specially constructed dividing engine. By means of these gratings he was able to measure the wavelengths of the different colors of light. Nobility was conferred on him in

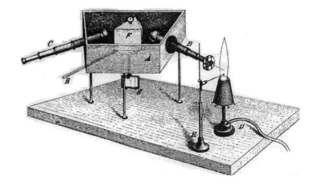

Fig. 6.3 The prototype spectroscope invented by Bunsen and Kirchhoff

1824 and shortly before his untimely death, he was made a knight of the Danish order of Danebrog [25].

It would be decades before the connection between Fraunhofer's lines and the absorption lines of metal atoms in the dark line spectrum would be made. It would be decades more before the dark line spectrum would be utilized fully, by way of Australia, in the analytical technique known today as atomic absorption spectroscopy [26].

Meanwhile back in 1822, Sir John Herschel (1792–1871), son of Sir William Herschel (1738–1822), while studying the visible spectra of colored flames, observed: "The colours thus contributed by different objects to flame afford in many cases a ready and neat way of detecting extremely minute quantities of them" [27]. This prescient remark would be taken up almost four decades later in Germany by Gustav Kirchhoff (1824–1887) and Robert Wilhelm Bunsen (1811–1899) who were making studies with the purpose of characterizing the colors of heated elements. It was Kirchhoff in 1859 who realized that the observed frequencies of the various elements' emission lines in their bright line spectra corresponded to the frequencies observed in Fraunhofer's dark line spectra [28]. He concluded that the dark lines were due to the absorption of the characteristic frequencies of the elements present in the cooler outer layers of the sun's atmosphere, and that these would be the same frequencies that these elements would emit when excited by an energy source such as a flame. A further conclusion was that each element should exhibit a line spectrum characteristic of that element, enabling not only chemical analysis, but analysis long distance—as much as eight light years away [29]!

They had done earlier studies of the characteristic colors of heated elements, and a burner that had been in use in Bunsen's laboratory since 1855 was ideal for this purpose since it gave a virtually colorless, soot-free flame of constant size [30]. In the summer of 1859, Kirchhoff suggested to Bunsen that they systematize their studies and try to develop a device that would form spectra of these colors by using a prism. By October of that year they had invented an appropriate instrument, a prototype spectroscope, shown in Fig. 6.3 [31].

Fig. 6.4 Robert Wilhelm
Bunsen (1811–1899),
renowned German chemist
and a founder of analytical
spectroscopy, who formed
generations of chemists and
physicists in his laboratory.
Photograph with the kind
permission of the Biblioteca
di Scienze dell'Università di
Firenze, Sezione Polo
Scientifico

Using this very first spectroscope, they were able to identify the characteristic spectra of sodium, lithium, and potassium. After numerous laborious purifications, Bunsen proved that highly pure samples gave unique spectra. In the course of this work, Bunsen detected previously unknown new blue spectral emission lines in samples of brine water from Bad Dürkheim and other well-known German spas. He hypothesized that these lines indicated the existence of a hitherto unknown element. After careful distillation of 600 quintals (about 44 tons) of this water, in the spring of 1860 he was able to isolate and discern two blue spectral lines in the several liters of residue, and recognized them as the signature of a new element. He named the element "cesium," after the Latin word for "deep blue". The following year he discovered rubidium by a similar process, or rather, one of his students recognized the existence of this new element in the same distillation residue. Spectroscopy had come a long way from looking at and contemplating the spectrum [32–34]. Figure 6.4 is an 1887 photograph of Bunsen.

Figure 6.5 is a photograph of a modern student spectroscope that has not departed very much from the original Bunsen-Kirchhoff model. The great difference is that this spectroscope uses as its dispersive element a replica grating made of polyethylene, an offshoot of the invention of David Rittenhouse (1732–1796) [35] who suspended 50 hairs between two finely divided screws, a method used 35 years later by Fraunhofer for his dark line spectral measurements. The digital camera in the foreground is connected to a computer screen (left) so that the spectrum of the light source, a helium lamp (right), can be seen on the screen. A built-in computer program can then measure the wavelengths of the helium spectral lines.

Just 1 year after the discovery of rubidium, in 1862, Anders Jonas Ångstrom (1814–1874) used the technique of thermonuclear fusion, i.e., observation of the

Fig. 6.5 A modern student
spectroscope. Photograph
courtesy of Richard Hermens
and MicroLab, Bozeman,
Montana

solar spectrum, to show that the sun's atmosphere contained hydrogen and to obtain
the spectrum of gaseous hydrogen distributed among many states and energy levels
[36]. In 1868, he published the results of a precise, systematic study of absolute
wavelengths of over 1,000 solar spectral lines, which he expressed in units of
10^{-10} m. This unit is now commonly known as the Ångstrom (Å). It was Ångstrom's
measurements that Johann Balmer used in 1885 to develop his empirical equation
for the series in the hydrogen spectrum that bears his name (please see Sect. 3.3).
And, of course, it was one of the key relationships leading to the proposal of the Bohr
atom 30 years later, with a stopover in quantum land in between.

At this point, one might ask just what more could spectroscopy accomplish? In
the same year that rubidium was discovered via spectroscopy, Sir William Crookes
(1832–1919) discovered thallium by the same method [37, 38], though credit is
also given to Claude-Auguste Lamy (1820–1878) who discovered it almost
simultaneously and working independently [39, 40]. It was Lamy who pointed out
the high toxicity of thallium and it compounds, a fact that was not lost on Agatha
Christie (See her novel "The Pale Horse") and the putative 21st century murderers
of former KGB officer Alexander Litvinenko.

And then came helium, the element named for the sun. It was the French
astronomer Pierre Jules César Janssen (1824–1907) who first recognized this new
element's yellow spectral line at 587.49 nm in the solar corona while observing a
total eclipse of the sun on August 18, 1868 in India. A few months later, on
October 20, the English astronomer Joseph Norman Lockyer (1836–1920) also
realized that the yellow line he had observed during the same eclipse (from
England) might be a new element—in collaboration with the chemist, Edward

Fig. 6.6 A simple
spectrometer. © 1998,
M.V. Orna

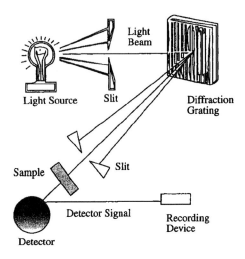

Frankland (1825–1899), he postulated the existence of a new element that they named helium (after the Greek word for sun, *helios*), the first element to be discovered extraterrestrially before its existence could be found on earth. All three receive credit for the discovery [41, 42]. In 1895, Sir William Ramsay (1852–1916) discovered helium on earth by examining the spectrum of an inert gas obtained by treating a uranium mineral with sulfuric acid; it was later found that the helium trapped in radioactive minerals were the alpha particles emitted in the decay process [43]. It was thought that helium would remain a rare element until the 1905–1906 discovery of David F. McFarland (1878–1955) and Hamilton P. Cady (1874–1943) of the chemistry department at the University of Kansas. They found that helium could be isolated from natural gas containing as much as almost 2 % of the element. When they published their complete findings in November of 1907 [44], Cady commented that their work "assures the fact that helium is no longer a rare element, but a common element, existing in goodly quantity for uses that are yet to be found for it" [45]. On April 15, 2000, the American Chemical Society declared the site of the discovery and analysis at the University of Kansas a National Historic Chemical Landmark.

From these remarkable beginnings, even more remarkable progress has been made. Emission spectroscopic methods, while actually the foundation stone of all spectroscopy, had a hard time making it into the industrial mainstream—it was only in the early 1930 s that industry began to seriously investigate the potential for optical emission spectroscopy as a technique to meet the demands of the industrial community [46]. Meanwhile the technique eventually evolved into absorption spectroscopic methods wherein one was not limited to viewing substances in flames or electric arcs, thus broadening the scope of qualitative analyte study to quantitation and to molecules that might otherwise be destroyed by employing the spectroscopic method. Ultraviolet-visible absorption spectrometers were developed utilizing both prisms and gratings as dispersion methods, set up in such a way that absorbance at a single wavelength (really a range of wavelengths

Fig. 6.7 Bausch and Lomb
Duboscq type colorimeter.
Photo by Gregory Tobias.
Courtesy of the Chemical
Heritage Foundation
Collections

depending on the slit width) or absorbance over the entire spectral range could be measured. Figure 6.6 illustrates a simple spectrometer that contains all the components necessary to conduct a spectrometric analysis. By varying the nature of the irradiating source, one can do measurements in the visible, ultraviolet and infrared regions of the spectrum. Rotation of the grating allows different wavelengths to enter the second slit and traverse the transparent sample.

The recording device could be any kind of electronic recorder or, in the early days, a human hand writing in a laboratory notebook.

Later spectroscopic developments such as infrared spectroscopy and nuclear magnetic resonance spectroscopy have been useful in characterizing and determining the structure of molecular species, and with respect to the interests of this work, of colorants [47]. Rapid advances in instrumentation have given rise to a plethora of spectrophotometric instruments coupled with monochromatic coherent light sources (lasers) and digital computers capable of not only allowing for very rapid determinations but also greater and greater instrument sensitivity and lower detection limits.

6.4 Color Measurement

The moment we broaden the purpose of visible spectroscopy to include areas beyond the discovery of new elements or the qualitative determination of already known elements, we perforce segue into the area of measuring color, as illustrated by the following declaration [22]:

August Beer, a German physicist and professor of mathematics at the University of Bonn, recognized the relationship between the absorption of light and concentration.

And of course, here we have the advent of Beer's Law, now studied by every introductory chemistry student—a law that lands us squarely in the land of colorimetry. It was actually Pierre Bouguer (1698–1758) who first discovered that the

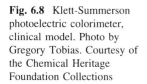

Fig. 6.8 Klett-Summerson photoelectric colorimeter, clinical model. Photo by Gregory Tobias. Courtesy of the Chemical Heritage Foundation Collections

quantity of sunlight passing through the atmosphere was inversely proportional to its thickness [48]. Johann Heinrich Lambert (1728–1777) had read Bouguer's *Essai* and quoted from it in 1760. Then August Beer (1825–1863) in 1852 extended the observation into a law that included the concentration of the absorbing material as well as its thickness [49]. Today it is known as the Bouguer–Beer–Lambert Law, the first law of colorimetry.

Based upon this law, color comparison instruments soon developed. One of the first instruments to use the absorption of light to determine concentration was the visual color comparator. The user visually compared the transmitted light from the sample and a standard solution and adjusted the path length until the transmitted light from both solutions appeared to have the same intensity [50]. Figure 6.7 shows an early color comparator called a Duboscq Colorimeter.

Shown in Fig. 6.8 is an early electronic Klett-Summerson colorimeter that was a "step above" the color comparator shown above. With this instrument the analyst could now abandon the unreliable human eye as detector and begin to rely on a variety of detection and signal recording methods. We have moved a long way from Duboscq and Klett, and yet these are the instruments that started us down the pathway of rapid, accurate, sensitive color measurement for a variety of purposes.

When thinking about making a measurement regarding color, we must distinguish between color measurement (a measurement that relates to what the observer sees) and a colorant measurement (a measurement that relates to the colorants used to color an object and that often does not relate to what the observer sees). Measuring color, then, comes down to three questions: am I measuring the color to define its color appearance, to determine if it matches another color, or to determine its composition? [51] Most of the time, chemists are concerned about the third question: the composition of a sample.

Early colorimeters measured (usually) light transmission at a single wavelength. A spectrophotometer measures color as energy distribution over a range of wavelengths. Although spectrophotometric techniques are not limited to the visible spectrum, most analyses of colorants are made in the visible region to simulate end use characteristics for assessment of quality. As pointed out earlier, UV and/or IR spectra are useful in characterizing the chemical structure of a colorant, but they contribute little to measurement of the subtle shade and strength differences that are important in industrial quality control.

At this point, we must distinguish between two types of "colorimetry". Measurement of color with the Klett-Summerson instrument shown above is measurement of the ratio of transmitted light to incident light set at a single wavelength as the light passes through a transparent sample. The reference is a blank, usually a sample tube filled with the solvent. This type of colorimetry is actually "single wavelength spectrophotometry". There is another type of colorimetry and other types of colorimeters that are used mainly in the industrial sector for purposes of color-matching against a set standard, usually a standard observer. This type of colorimetry is a valuable means of extending spectrophotometry because it provides dimensions for color specification and the detection of small shade differences between samples. Computers make it possible to benefit from both measurement techniques, for colorimetric data can be computed from spectrophotometric data [52].

To summarize, let us look at some of these terms again. Spectroscopy means to "look at the spectrum," but not necessarily the visible, nor even the electromagnetic, spectrum. There are emission, absorption, visible, ultraviolet, infrared, nuclear magnetic resonance, and mass spectroscopies—this is the broadest term of all and this is not an exhaustive list. Spectrometry means literally to "measure the spectrum," and we can apply the word to all the types of spectroscopies named above. Spectrophotometry means literally to "measure the spectrum with photons," and so can only include those types of spectrometries that utilize photons in this term. Even more particular is colorimetry, which means literally to "measure color," a term that can be ambiguous because it includes (1) a form of spectrophotometry and (2) a color measurement system based on color primaries for color-matching.

One last term we can introduce is "coloristics". This term is a catch-all general concept that can include all of the above and much more: everything that has to do with color with respect to its visualization, perception and measurement. Coloristics includes but is not limited to the following areas: color perception, color appearance, color order systems, colorimetry, color difference evaluation and standardization, application of colorimetry and color recipe calculation in different industries, the world of colored products and color in life style, color in environmental planning, chemical and technological research and development in different areas of producing colored products: dyes, pigments and additives, their production and testing in fields such as textile, leather, paper, food, paint, printing ink as well as plastics industries, etc. In other words, coloristics includes virtually everything that has to do with color. Color is universal.

In the next chapter we will look at just how universal color is. We will venture into the world of biological color—from a chemical viewpoint, of course.

References

1. Bukovsky V (1997) Yellowing of newspaper after deacidification with methyl magnesium carbonate. Restaurator Int J Preserv Libr Arch Mater 18(1):25–38
2. Gay-Lussac JL (1824) Instruction sur l'essai du chlorure de chaux. Ann Chim Phys 26:162–175
3. Scott JM (1950) Karl Friedrich Mohr (1806–1879) father of volumetric analysis. Chymia 3:191–203
4. Mohr KF (1855) Lehrbuch der chemisch-analytischen Titrirmethode nach eigenen Versuchen und systematisch dargestellt : für Chemiker, Ärzte und Pharmaceuten, Berg-und Hüttenmänner, Fabrikanten, Agronomen, Metallurgen, Münzbeamte etc.; in zwei Abtheilungen. Vieweg und Sohn Braunschweig, pp 73–82
5. Ihde AJ (1964) The development of modern chemistry. Dover, New York 291
6. Rubinson KA (1987) Chemical analysis. Little, Brown and Company, New York, pp 269–292
7. von Baeyer A (1871) Über die phenolfarbstoffe. Ber 4:658–659
8. Kekulé FA, Barbaglia G (1872) Action of o-sulfobenzoic anhydride or of o-sulfobenzoyl chloride on phenol. Ber 5:876
9. Mendes AJ (1999) A colorful ion-exchange experiment. J Chem Educ 76:1538–1540
10. Pantaleão I, Portugal AF, Mendes A, Gabriel J (2010) Carbon dioxide absorption in a membrane contactor with color change. J Chem Educ 87:1377–1379
11. Merzlyak MN, Gitelson A (1995) Why and what for the leaves are yellow in autumn? On the interpretation of optical spectra of senescing leaves (Acer platanoides L.). J Plant Physiol 145(3):315–320
12. Matile P, Flach BMP, Eller BM (1992) Autumn leaves of Ginkgo biloba L.: optical properties, pigments and optical brighteners. Bot Acta 105(1):13–17
13. Ihde AJ (1964b) The development of modern chemistry. Dover, New York, pp 571–572
14. Tsvet M (1906) Physikalisch-chemische Studien über das Chlorophyll. Die Adsorptionen. Ber Dtsch Bot Ges 24(316–323):321–322
15. Tsvet M (1906) Adsorption analysis and chromatographic method. Application to the chemistry of chlorophyll. Ber Dtsch Bot Ges 24:384–393
16. Barnes Z (1993) 1915 Nobel laureate Richard Martin Willstätter (1872–1942). In James LK (ed) Nobel laureates in chemistry: 1901–1992. American Chemical Society, Washington and Chemical Heritage Foundation, Philadelphia, pp 108–113
17. Sourkes TL (2009) The discovery and early history of carotene. Bull Hist Chem 34(1):32–38
18. Shetty PH (1993) 1952 Nobel laureate Archer John Porter Martin (1910). In James LK (ed) Nobel laureates in chemistry: 1901–1992. American Chemical Society, Washington and Chemical Heritage Foundation, Philadelphia, pp 352–355
19. Shetty PH (1993) 1952 Nobel laureate Richard Laurence Millington Synge (1914). In James LK (ed) Nobel laureates in chemistry: 1901–1992. American Chemical Society, Washington and Chemical Heritage Foundation, Philadelphia, pp 356–358
20. Stahl GA (1977) Interview with Archer J.P. Martin. J Chem Educ 54:80–83
21. Wollaston WH (1802) A method of examining refractive and dispersive powers, by prismatic reflection. Philos Trans R Soc 92:365–370
22. Thomas NC (1991) The early history of spectroscopy. J Chem Educ 68:631–634, p 632. Reprinted with permission from American Chemical Society © 1991
23. Fraunhofer J (1817) Bestimmung des Brechungs-und Farbenzerstreuungs-Vermögens verschiedener Glasarten, in Bezug auf die Vervollkommnung achromatischer Fernrohre. Gilbert's Annalen 56:264–313
24. Jaecks DH (1986) Developments in 18th century optics and early instrumentation. In: Stock JT, Orna MV (eds) The history and preservation of chemical instrumentation. D. Reidel Publishing Company, Dordrecht, pp 51–65
25. Merz S (1865) Das Leben und Wirken. Jos. Thomann'schen Buchdruckerei, Landshut

26. Walsh A (1955) The application of atomic absorption spectra to chemical analysis. Spectrochim Acta 7:108–117
27. Lewis J (1936) Spectroscopy in science and industry. Blackie and Son, Glasgow 4
28. Kirchhoff G (1860) Über das Verhältniss zwischen dem Emissionsvermögen und dem Absorptionsvermögen der Körper für Wärme und Licht. Ann Phys 185(2):275–301
29. Kirchhoff G (1860) Über die Fraunhofer'schen Linien. Ann Phys 185(1):148–150
30. Jensen WB (2005) The origin of the Bunsen burner. J Chem Educ 82:518
31. Kirchhoff G, Bunsen R (1860) I. Chemische Analyse durch Spectralbeobachtungen (Chemical analysis by observation of spectra). Annalen der Physik und der Chemie (Poggendorff) 110:161–189, pp 162–163
32. Kirchhoff G, Bunsen R (1861) Chemische Analyse durch Spectralbeobachtungen. Annalen der Physik und Chemie 189(7):337–381
33. Weeks ME, Leicester HM (1968) Discovery of the elements, 7th edn, Journal of Chemical Education. Easton, Pennsylvania, pp 598–603
34. Weeks ME (1932) The discovery of the elements. XIII. Some spectroscopic discoveries. J Chem Educ 9:1413–1434
35. Cope T (1932) The Rittenhouse diffraction grating. J Franklin Inst 214:99–104. Historical note: Although the invention of the diffraction grating is often attributed to Joseph Fraunhofer in 1821, thirty-five years earlier David Rittenhouse read a paper before the American Philosophical Society on 17 February 1786 (Trans Am Philos Soc, Vol. 2) in which he described his construction of what we now call a diffraction grating, together with his observations of its effects: an array of spectra in six orders with red light deviated the most and blue light the least. He also reported a law of distribution of these spectra in space, a corollary of the more general law discovered thirty years later by Augustin Jean Fresnel (1788–1827) and still later, independently, by Fraunhofer
36. Ångström AJ (1862) Über die Fraunhofer'schen Linien im Sonnenspectrum. Ann Phys 193:290–302
37. Crookes W (1861) On the existence of a new element, probably of the sulphur group. Chem News 3:193–194
38. Crookes W (1862) Preliminary researches on thallium. Proc R Soc Lond 12:150–159
39. Lamy C-A (1862) De l'existencè d'un nouveau métal, le thallium. Comptes Rendus 54:1255–1262
40. Weeks ME (1932) Discovery of the elements XIII. Supplementary note on the discovery of thallium. J Chem Educ 9:2078
41. Weeks ME, Leicester HM (1968) Discovery of the elements, 7th edn, Journal of Chemical Education. Easton, Pennsylvania, pp 757–767
42. Lockyer N (1928) The story of helium. In: Lockyer TM, Lockyer WL (eds) Life and works of Sir Norman Lockyer. MacMillan and Company, London, pp 266–291
43. Hunter NW, Zeigler K (1993) 1904 Nobel laureate William Ramsay 1852–1916. In James LK (ed) Nobel laureates in chemistry 1901–1992. American Chemical Society, Washington and Chemical Heritage Foundation, Philadelphia, pp 23–29
44. Cady HP, McFarland DF (1907) The occurrence of helium in natural gas and the composition of natural gas. J Am Chem Soc 29:1523–1536
45. Siebel CW (1968) Helium, child of the sun. The University Press of Kansas, Lawrence
46. Angelotti WJ (1986) The history of optical emission techniques for the industrial user. In: Stock JT, Orna MV (eds) The history and preservation of chemical instrumentation. D. Reidel Publishing Co., Dordrecht, pp 67–78
47. Venkataraman K (1977) The analytical chemistry of synthetic dyes. Wiley-Interscience, New York
48. Bouguer P (1729) Essai sur la gradation de la lumière. Section II. De la transparence et de l'opacité. II. De la proportion selon laquelle la lumière diminué en traversant les milieux. Chez Claude Jombert, Paris, p 44
49. Beer A (1852) Bestimmung der Absorption des roten Lichts in färbigen Flüssigkeiten. Ann Phys Chem 86:78–90

50. Altemose IR (1986) Evolution of instrumentation for UV-visible spectrophotometry. Part I.
    J Chem Educ 63:A216–A223
51. Berns RS (2000) Billmeyer and Saltzman's principles of color technology, 3rd edn. Wiley,
    New York 76
52. Garland CE (1977) Solution coloristics. In: Venkataraman K (ed) The analytical chemistry of
    synthetic dyes. Wiley-Interscience, New York, pp 149–171

Chapter 7
Color on the Biological and Biochemical Front

One of the many things that the German dye companies excelled at was seeing the long-range and global picture and acting upon it. They went very large-scale on virtually every front, utilizing all the weapons that science and technology had to offer. They found that doing so enabled them to synthesize two important natural colorants, indigo and alizarin, thus eliminating the industry's dependence on natural products and imports—the dye industry could now not only be self-sufficient, but could use the know-how generated in these key syntheses to make many other derivatives. In addition to setting up large laboratories that employed hundreds of chemists, they built libraries that had subscriptions to virtually every scholarly science journal in the Western world, they established dedicated bureaus to patent discoveries, and they virtually "wrote the book" on abstracting services by having these offices work around the clock and not only in the area of chemistry, but in every other domain that might afford them related knowledge—biology, the emerging area of biochemistry, pharmaceuticals, photography, and explosives [1].

Such a wide range of interests coupled with a prodigious outpouring of time and money would never have been successful unless there was a great deal of scientific talent to accompany these resources. And talent is often driven by an insatiable curiosity to know: this was the characteristic of much nineteenth century research on biological systems—much of it centered around the phenomenon of color.

7.1 Reduction and Oxidation of the Vat Dyes

In 1809, Michel Chevreul (1786–1889) described the reactions of indigo blue (*Indigofera tinctoria*) and "Pastel" (*Isatis tinctoria*), not realizing that the blue coloring matter was the same chemical compound [2]. He carried out a reduction of the blue color to yield an "indigo white," and there followed extensive studies of its chemistry by others. Walter Crum (1796–1867) purified indigo blue by sublimation and Auguste Laurent (1807–1853) laid the foundations for the

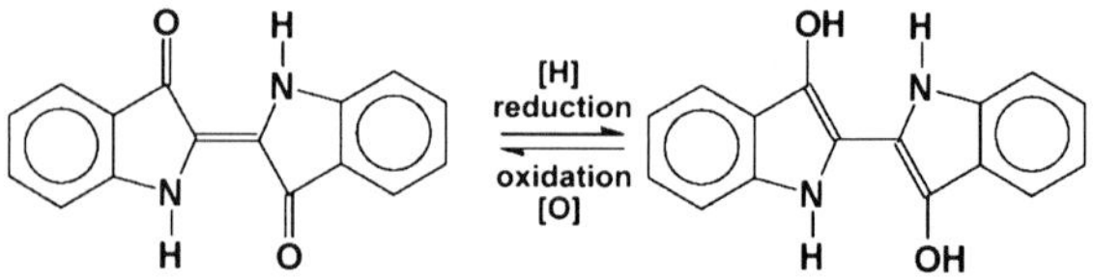

Fig. 7.1 Reduction of the blue form of indigo (*left*) to near colorless leucoindigo (*right*)

subsequent elucidation of its structure by Adolf von Baeyer (1835–1917). The synthesis and chemical and physical properties of indigo are described in Sect. 5.5. The actual reversible reduction to the leuco form is shown in Fig. 7.1.

Leucoindigo is a white crystalline sparingly-soluble solid that is formed by reduction of water-insoluble indigo blue by a variety of reducing agents in the laboratory, and historically by biochemical reduction in a fermentation vat, a process that is the basis for vat dyeing [3]. In the presence of an excess of alkali, such as NaOH, the acid is converted to the water-soluble mono- and/or di-anion leuco salt to produce a yellow solution. It is into this reduced solution that a textile is immersed, and after removal into the atmosphere, is then air-oxidized back into the original insoluble indigo pigment. The reduction of natural pigments like indigo and the reoxidation of the reduced form were important reactions in the pioneer studies of Christian Friedrich Schönbein (1799–1868) and of Moritz Traube (1826–1894) on biological oxidation [4]. This work on reversible color reactions would soon be applied to studies on human body fluids, particularly the blood.

7.2 Research on the Color of Blood

Jöns Jacob Berzelius (1779–1848) named the blood pigment hematin, although this term now refers more specifically to a component of the pigment and not of the pigment itself. The present medical definition is "a blue to blackish-brown compound formed in the oxidation of hemoglobin and containing ferric iron (also called ferriheme, oxyheme, oxyhemochromogen)." The derivation, of course is from *haimatos*, Greek for blood, and this prefix has carried over into other contexts such as hematite, a bloodlike stone, hemophilia, hemorrhage, and other terms that are associated with blood. In 1835, Brett and Goldincj Bird treated blood with acid and examined the products from the reaction [5]. In 1838 Lecanu (1800–1871) established the association of the blood iron with the pigment that Berzelius had named [6], but it was not clear in the 1850s if this association involved the red blood cells. The reasons for the difference in the color of arterial and venous blood was unclear at that time. Bernard (1813–1878) [7] described an extensive series of physiological experiments which led eventually to recognition that the color change was due to oxygenation and deoxygenation of the blood hemoglobin (Hb) through an easily dissociable mechanism of Hb-O_2 versus Hb-CO_2. Hoppe-Seyler (1825–1895) [8] and Stokes (1819–1903) did spectroscopic work that eventually led to this conclusion [9]:

...the colouring matter of blood, like indigo, is capable of existing in two states of oxidation, distinguishable by a difference in colour and a fundamental difference in the action on the spectrum. It may be made to pass from the more to the less oxidized state by the action of suitable reducing agents, and recovers its oxygen by absorption from the air...

Stokes also performed spectroscopic studies on blood, finding that treatment with acid caused a profound change in the visible spectrum. Later research revealed the complex participation of proteins and molecular fragments, products of the denaturation of the globin portion of hemoglobin.

7.3 The Towering Figure of the Multifaceted Paul Ehrlich

7.3.1 Ehrlich's Early Colorful Work with Histological Staining

In 1877, Ehrlich (1854–1915) reported the first systematic study of biological staining properties, work that followed directly from the appearance and synthesis of the aniline dyes after Perkin's discovery of mauve in 1856. In the decades following the advent of mauve, chemists and biologists quickly discovered that dyes could stain cellular tissue. von Gerlach's 1858 seminal book on the subject [10] kicked off a flurry of experiments such that histological staining became almost commonplace [11]. Ehrlich found that the dyes which were basic in nature, such as methylene blue, had the capacity of deeply staining cell nuclei and he coined the word "basophilic" for the cellular elements that took up basic dyes. On the other hand, the cytoplasm seemed to be immune to basic dye uptake, but could be preferentially stained by dyes that were acidic in character such as eosin or fuchsin; Ehrlich called these cell elements "oxyphiles." [12] This activity was corroborated by Zacharias (1852–1911) who digested away the nuclear cytoplasm and isolated nuclei that were readily stained by suitable dyes [13]. Flemming (1843–1905), the nineteenth century biologist, who gave the first accurate description of nuclear division and named it "mitosis," followed up on this work by observing that [14]:

the framework [of the cell nucleus] owes its refractile character, the nature of its reactions, and particularly its affinity for dyes to a substance that I have tentatively named chromatin because of the last-named property...I retain the word chromatin until a decision about it is made by chemical means, and I use it entirely empirically to denote the substance in the cell nucleus which is stained by nuclear dyes.

Flemming's tentative term for the cell material that takes up dyes, chromatin, was enlarged upon by Wilhelm von Waldeyer-Hartz (1841–1923) who, in 1888, called the threadlike structures in the nucleus "chromosomes," or colored bodies. Thus our present-day vocabulary with which we describe cellular components had its origin in color reactions, and remains a colorful reminder of that colorful past.

7.3.2 Ehrlich's Major Ideas

Meanwhile back to Ehrlich who achieved renown as an organic chemist, histologist, immunologist, hematologist and pharmacologist. It will be worthwhile to follow the thought processes of someone with such great accomplishments in so many related biomedical fields: immunology, cancer research, and chemotherapy. Ehrlich disliked the formality of school but managed to excel in Latin and mathematics. As a medical student, Ehrlich was captivated by structural organic chemistry and dyes. When he was 23, his first paper was published on selective staining. This seminal paper was the precursor of his doctoral thesis, "Contribution to the Theory and Practice of Histological Staining" which contained most of the major ideas that would guide his future career [15].

What were some of these major ideas? Ehrlich in his dissertation noted that the mechanism of the binding of dyes to tissues was more than merely physical adhesion but suggested a chemical interaction; non-stoichiometric but nevertheless chemical in nature. He also recognized that the chemical group on the dye molecule responsible for its color was generally not the same as the group responsible for the molecule's affinity for the tissues [16]. This observation eventually led to his famous "side-chain theory" of antibody formation which led to his being awarded the 1908 Nobel Prize in Medicine; it also was the germ of his ideas on chemoreceptors, a concept that eventually led to the theoretical and practical basis of chemotherapy [17].

Ehrlich's biographer observes that the theme of different functionalities on different parts of the same molecule runs throughout his work [18]. Another example that Ehrlich gave is the interaction of light with a molecule. He observed that [19]:

> light emission and absorption are not a function of the whole molecule...the light absorption of indigo is unchanged if it is converted into indigo-blue-sulfonic acid; we can therefore conclude that the atom group which absorbs the yellow light so strongly and produces the blue color of indigo is not the same as unites with the sulfonic acid, but another....

7.3.3 The Birth of Chemotherapy

Another great idea arose from Ehrlich's own experimental work on the affinities of various dyes for specific tissues. In 1886, doing "vital staining" with the dye methylene blue led him to consideration of "localized organ therapy." In other words, he discovered that the methylene blue had a special affinity for nerve cells and wondered if similar selectivity on the part of other dyes for other types of cells might exist. In particular, if a dye were to have an affinity for a disease-causing organism, then it might also be possible to target that organism. He took this idea a step further: if a toxin acted in a similar manner as the dye, then one could deliver a toxin to that organism along with the agent of selectivity [20]. Using this

Fig. 7.2 Portrait of Paul
Ehrlich, 20th Century.
H. Hinkley. Oil on canvas
mounted on masonite. Gift of
Fisher Scientific
International. Chemical
Heritage Foundation
Collections. Photograph by
Will Brown

principle, he successfully treated certain experimental trypanosomal infections
with azo dyes [11, 15].

The fixing of the dyes could be by either of two mechanisms, the first like lake
formation—the combination of the dye with a constituent of the fabric to form an
insoluble salt-like compound—thus immobilizing or localizing the drug. The
second, the formation of solid solutions, where the dye forms a homogeneous
mixture with the substance of the fabric. He suggested that certain drugs might be
fixed in cells through a similar process [21]. Figure 7.2 is a portrait of Ehrlich from
the Fisher Collection, Chemical Heritage Foundation.

In his earlier work he made no mention of a specific chemical combination
between the drug and a receptor or side-chain in the cell. However, he changed his
mind in 1913 due to the work of Langley (1852–1925) on receptors [17, 22] and
his own work on drug resistance [23]:

> *For many reasons I had hesitated to apply these ideas about receptors to chemical
> substances in general, and in this connection, it was, in particular, the brilliant investi-
> gations by Langley, on the effects of alkaloids, which caused my doubts to disappear and
> made the existence of chemoreceptors seem probable to me.*

The theory formed the theoretical basis for his work on chemotherapy, and the
technique became popularly known as the "magic bullet" technique, the use of a
drug that would kill only the agent being targeted.

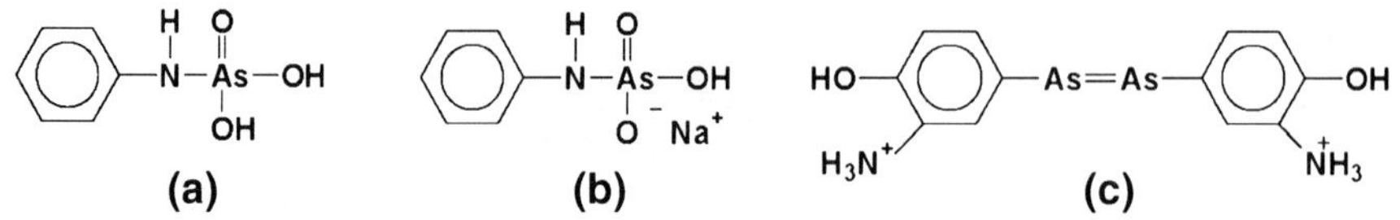

Fig. 7.3 Arsenic organoderivatives that figured in the development of Salvarsan

7.3.4 The Salvarsan Story

Although in the process of using dyes as "magic bullets" Ehrlich made lasting contributions to the art and science of histological staining, gradually the emphasis moved away from dyes to any type of compound that could exhibit tissue selectivity. Using this great idea and building upon the work of others, Ehrlich looked far afield for what he needed.

In 1863 at the University of Strasbourg, Pierre Jacques Antoine Béchamp (1816–1908) prepared an arsenic compound by heating aniline with aniline arsenite [24] that entered the literature as "Béchamp's anilide" or the anilide of arsonic acid (Fig. 7.3a).

Ehrlich somehow became aware of this compound—perhaps because he too had spent some time at Strasbourg and was familiar with the work and the literature being published there, perhaps because he had learned of this compound's antimicrobial activity against *Trypanosoma* in a form called "Atoxyl" because it was less toxic than other arsenic compounds—and set to work to characterize it. He first found out that it was not the anilide at all but actually 4-aminophenylarsonic acid [25], and that "Atoxyl" was its sodium salt (Fig. 7.3b). Based on Atoxyl's activity, Ehrlich was encouraged to seek an antisyphilitic (that is, a compound that would specifically target the spirochete that caused syphilis, *Treponema pallidum*) among arsenic organoderivatives. He and his co-workers Sahachiro Hata and Bertheim, after systematically screening hundreds of newly synthesized arsenic compounds, eventually found arsphenamine. Ehrlich postulated that it had the structure shown in Fig. 7.3c (marketed later by Hoechst AG as *Salvarsan*), the first effective treatment against syphilis, the first modern chemopherapeutic compound, and Ehrlich's crowning achievement [26, 27]. *Salvarsan* would eventually become the most highly prescribed drug in the world [1].

Salvarsan was also called by another name that designated its order of discovery in the systematic screening train in Ehrlich's laboratory. It is Compound 606—erroneously thought to be the 606th compound tested, but really the 6th compound in the 6th group of chemicals being tested—not quite 606 compounds but a prodigious amount of work nonetheless. A curious thing about *Salvarsan* is that it is not a chemical compound at all, but a mixture of compounds. It was found that there was no evidence for the arsenic-arsenic double bond assignment in Fig. 7.3c [27, 28] initially attributed to it by Ehrlich himself. His synthetic method was not entirely reproducible, which no doubt contributed to the varying toxicity of different Salvarsan batches, nor was the arsenic-arsenic double bond plausible

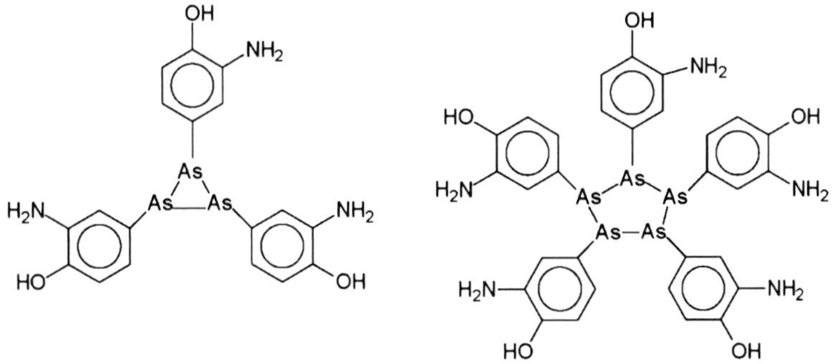

Fig. 7.4 The chemical characterization of Salvarsan as reported in 2005 by a team of chemists working at the University of Waikato, New Zealand

as it is found normally only in sterically crowded molecules. Thus the true structure of Salvarsan was open to question and debate for almost a century. It was not until 2005 that a group working in New Zealand reported the first definitive evidence for the composition of Salvarsan based on electrospray ionization mass spectrometric data [27] and found it to be a mixture of cyclic arsenic species, $(RAs)_n$, with n = 3 and n = 5 as shown in Fig. 7.4. The New Zealand chemists also report that the mixture of cyclic $(RAs)_n$ compounds slowly release $RAs(OH)_2$ on oxidation, presumably giving rise to Salvarsan's antisyphilitic properties.

7.3.5 A Summary of Paul Ehrlich's Accomplishments from the Standpoint of Dyes and Colors

Ehrlich was an innately intelligent man who was not afraid to learn from others and not afraid to change his mind, as has been amply documented in some of the vignettes above. He was also a keen observer who thought about what he observed, asking for the chemical reasons behind the observation, and then devising brilliant and novel experimental methods for verification. He was also a very hard worker, spending almost every waking minute at his work in the laboratory. Although his area of work was mainly biochemical/biomedical, he never strayed from the chemical principles of structure and bonding—he could even be called the father of molecular biology. Listed below are some of his accomplishments and qualities as gleaned from some key sources [15, 17, 20, 21, 29].

- Dyes for the most part adhere to animal fibers without the need of a mordant since animal fibers normally contain nitrogen with a lone pair of electrons affording a foothold for the dye—this led Ehrlich to try staining animal tissues using those same dyes
- Colored molecules were his labeling, measuring, and diagnostic tools
- He used the products of the dye industry to argue for a new, chemical approach to biomedical research

- In his doctoral thesis of 1878 he outlined and classified the major synthetic dyes, resolved problems of commercial names and contaminants, and described how dyes could be used as staining agents to differentiate tissues
- He classified dyes into three groups: the primary amino dyes, the sulfonic acid derivatives of aniline blue, and the acidic dyes containing nitro and halogen substituents
- He learned how to exploit the loss of color as dyes were reduced to develop a tool to enable semi-quantitative measurements and provide information about cell surfaces—molecules of dyestuffs in his hands became probes and measuring devices
- Ehrlich learned to exploit the wide-ranging responses of vat dyes to reducing agents with two criteria: ease of reducibility and insolubility
- He was aware of and pointed out Perkin's dictum that the property of color and the ability to impart color were not the same thing
- His tight connections with industry enabled him to transform commercial chemical products into analytical tools in the service of biological investigations
- His early grasp of the versatility and potential use of the new dyestuffs technology came shortly before a time of increasing interest both in public health services, as a means of ensuring social stability, and in the conquest of tropical diseases
- He had the ability to think independently, and to pursue unconventional thinking about dye interaction with cells, convinced that nature is unreliable in its adaptive reactions to pathological insults.

One can cite many more reasons for Ehrlich's brilliant success in virtually every field. All of them are necessary, but not sufficient, to account for his monumental achievements. But there is one more ingredient that has not been mentioned and is absolutely necessary: Ehrlich was passionate about his subject. He was intrigued, he was curious, he could not delve deeply enough, in short, he was a man in love.

7.3.6 Paul Ehrlich: The Person

Ehrlich was born in Strehlin, Germany (presently Strzelin, Poland, after the national boundary shifts of two world wars) to Ismar Ehrlich and Rosa Weigert Ehrlich, the fourth of five children, and only son, in this tight-knit Jewish family. Although he disliked taking examinations and was considered a mediocre student, he exhibited an early interest in things scientific: he even instructed the town pharmacist on how to prepare cough drops according to his own formula [18]. His upbringing had a telling influence on his thinking and behavior for the rest of his life, as is evidenced from one of his letters in which he claims that the oppressive burden of school curtailed his lifelong zest for freedom [30].

This zest for freedom carried through to his university studies where, for the most part, he did not attend the formal classes but preferred to read those subjects that were of interest to him—even eschewing attendance at the lectures of one of Strasbourg's most famous chemists and later father of the German dye industry, Adolf von Baeyer (1835–1917). Ehrlich had certain scientific goals and seemed to concentrate on learning only those ideas that would directly fulfill his intellectual targets [14]:

> *One wonders what direction Ehrlich's future career would have taken if he had come under the influence of von Baeyer, the chemist, rather than [Wilhelm] Waldeyer, the histologist. Ehrlich already realized that he had an instinctive talent for visualizing structural formulae in three dimensions. It was this ability to see benzene rings and their side-chains in stereoscopic view in his mind's eye that was of supreme value in his later research...Ehrlich chose medicine as a profession, but his real passions were organic chemistry and histology.*

His great passions were organic chemistry and histology, having great talent with the respect to the former in that he could visualize organic structures in three dimensions. Yet he chose medicine as a profession.

Later in his career, Ehrlich was appointed director of a "facility" that limped along on a skimpy budget. It had the pretentious title of the "Institute for Serum Research and Serum Testing," but the working conditions there prompted him to remark to his friend, August von Wassermann, as he showed him around, that all he needed were test tubes, a Bunsen burner, and blotting paper, and that he could even work in a barn if necessary [31]. Living proof that great work can be accomplished without the multimillion dollar budgets now common on the scientific scene!

7.4 Some Other Biochemical Spin-Offs from the German Dye Industry and Paul Ehrlich's Legacy

The world of Nobel laureates seems to be peppered with "refugees" from the dye industry, for one reason or another. Two of them, a 1939 and a 1948 laureate, both in physiology or medicine, deserve to be mentioned in this context. One of them, Domagk had to wait 8 years to finally accept his award. He had the misfortune to be a German national at the height of Hitler's power and the Nazi government forced him to sign a letter renouncing the prize. The other, Müller, was lucky enough to be Swiss and a post-Hitler laureate, and thus traveled to Stockholm on time to collect his prize. And many years later, the scientific trio of Köhler, Milstein, and Jerne, still wearing Ehrlich's chemotherapeutical mantle, walked off with the 1984 Nobel Prize in Physiology or Medicine for their work on monoclonal antibodies.

7.4.1 Gerhard Johannes Domagk (1895–1964)

Domagk saw service at the front in World War I, where he was deeply affected by the plight of so many soldiers and the medical teams caring for them: they were utterly helpless against the bacterial infections that actually did worse damage than the wounds themselves, many soldiers dying within days of medical intervention. After the war, he returned to school and received his medical degree. In 1924 he took the post of University Lecturer in Pathological Anatomy at the University of Münster, but later left to do research in industry. And it was there in 1932 that he would accomplish his lifetime major achievement. The site was the new research institute set up by the I.G. Farbenindustrie, a major player in the German dye industry, at Wuppertal. Domagk took up almost, as it were, where Ehrlich had left off: finding antibiotic uses for the dyes produced by the company. At first meeting with little success, he and his team tried an old chemical trick used to make dyes bind better to wool: he introduced a sulfonamide function into azo dyes and found that one of the new derivatives successfully cured mice infected with an extremely virulent strain of *Streptococcus pyogenes*. The new substance, dubbed *Prontosil rubrum*, and other sulfonamides, quickly became virtually the only weapons physicians had against bacterial infections until the mass production of penicillin became a reality. For this discovery, originally from a dye, Domagk was awarded the 1939 Nobel Prize in Medicine. He finally received the prize in 1947 after the demise of the Third Reich [32, 33].

7.4.2 Paul Hermann Müller (1899–1965)

Another dye chemist working in the first half of the twentieth century, this time at the J. R. Geigy Company, Müller, moved out of the dye area when he found that the fumes gave him asthma. He transferred into research but used his knowledge of dyes to produce light-resistant materials and to study the stability of compounds. He eventually entered the area of disinfectants and the search for effective insecticides—a search that eventually led to the production of DDT (dichlorodiphenyltrichloroethane). Although Müller was not the first to synthesize the compound, it is he who discovered its insecticidal effects and for this he was awarded the 1948 Nobel Prize in Medicine [34]. And DDT has been effective indeed. Its use virtually wiped out the incidence of malaria in many parts of the world, but as with most other boons, it has its disadvantageous side. Signaled by Rachel Carson in 1962 as a major cause of impeded bird reproduction, and later by the fact that it remains toxic in the soil with as long as a 30-year half-life, DDT was banned by the United States in 1972. Ever controversial, its proponents claim that many of the newer agents developed to replace it are not effective enough and are too expensive. It is presently lawful to use it in many parts of the world for Indoor Residual Spraying (IRS) where its more harmful effects on the environment would not be at issue [35].

7.4.3 Monoclonal Antibodies

Early in the twentieth century, Ehrlich, from his tissue staining experience, had postulated that if a dye could be made that directly targeted a specific disease-causing organism that perhaps a toxin could be made to accompany the dye and selectively destroy the organism. This idea, dubbed early on as a "magic bullet" mechanism, was later expanded beyond colored molecules to include any type of molecular toxin delivery system. But Ehrlich's dream was far from being realized as no molecular species was selective enough at the cellular level: this had to wait for an understanding of the antibody-antigen relationship, the technique of cell cloning, and the fusing of different types of cells in the hopes of bringing about the needed antibody specificity.

An antigen is an **anti**body **gen**erator, from which the word is derived. An antigen, usually protein in nature, is signaled as a foreign body to a given organism, thus triggering a response on the part of the immune system: the generation of an antibody to combat the invader. A monoclonal antibody is one produced by a single clone of cells, that is, a single cell and its progeny. But what was yet needed was the ability to produce identical antibodies specific to a given antigen, a technique that was finally developed, but not without controversy [36], by Cambridge University's Milstein (1927–2002) and a postdoctoral fellow working in his laboratory, Köhler (1946–1995) [37]. Jerne (1911–1994), working out of the Basel Institute for Immunology, developed the theory that buttressed the work of Milstein and Köhler [38]. The three shared the Nobel Prize for Physiology or Medicine in 1984 for the discovery.

Although the development of monoclonal antibody technology may seem a long step from the vision of a student Ehrlich, face and hands covered with biological stains, this cornerstone of modern immunochemistry may never have happened if it were not for Ehrlich's dream—a dream now come true [39]:

> *[Monoclonal antibodies have] provided an enormous opportunity for examination of a range of previously elusive issues. For example, [they] are being used in radioimmuno-assays [RIA], enzyme-linked immunosorbent assays [ELISA], immunocytopathology, and flow cytometry for invitro diagnosis and immunotherapy of human disease…[they] are just beginning to fulfill that promise inherent in their great specificity for recognizing and selectively binding to antigens on cells.* [From Waldmann (1991) Monoclonal antibodies in diagnosis and therapy. Science 252:1657–1662; p. 1657; reprinted with permission from AAAS.]

7.5 More Colorful Natural Products Chemistry

7.5.1 Chlorophyll

Natural products exhibiting a range of spectral colors have already been covered in various parts of this book. Red and blue, exemplified by alizarin and indigo respectively, have been shown to have been the cornerstones of the development of the dye industry. But it was green and yellow that caught of eye of chemists like Willstätter and led to some basic understandings of how chlorophyll works in transforming the sun's energy into plant substance. Major steps along the way are tabulated in Table 7.1, focusing largely on the isolation and characterization of the chemical substances involved.

Table 7.1 Milestones in chlorophyll research

Date	Scientist(s)	Accomplishment	Remarks	References
1640s	Jan Baptista van Helmont (1577–1644)	Conducted quantitative experiment on weight gain by a tree; concluded erroneously that gain was due to water	The experiment was far ahead of its time, 100 years before the formulation of the law of conservation of matter	[40]
1779	Jan Ingenhousz (1730–1799)	Discovered the role of light in the photosynthetic process	Summarized photosynthesis in terms of the new chemistry of Lavoisier	[40]
1818	Pelletier (1788–1842) and Caventou (1795–1887)	Isolation (in impure form) and naming of chlorophyll	Also commemorated for isolation of quinine from chinchona bark (See Fig. 7.5)	[41]
1837–1838	J. J. Berzelius (1779–1848)	First chemical investigation of leaf pigments	Proposed that chlorophyll was a mixture of at least two compounds	[42]
1866	Hoppe-Seyler (1825–1895)	Recognized relationship between blood and leaf pigments	Close structural relationship of their molecules to one another	[43]
1906	M. S. Tsvet (1872–1919)	Separation of chlorophylls by chromatography	Proved that chlorophyll is a mixture of two green and several yellow pigments	[44]
1906	Alexandre Léon Etard (1852–1910)	Showed the existence of a large series of different chlorophylls	Some of these "chlorophylls" were not distinguished from waxes	[43]
1912–1914	Willstätter (1872–1942)	Fundamental structural studies on chlorophyll (and its empirical formula) and other plant pigments	Recognition of central metal ion of Mg bound to four nitrogens; 1915 Nobel Prize in Chemistry	[43]
1930	Fischer (1881–1945)	Determination of complete structure of chlorophyll	1930 Nobel Prize in Chemistry	[45]
Late 1940s	Woodward (1917–1979)	Complete synthesis of chlorophyll (along with a myriad of other complex molecules)	1965 Nobel Prize in Chemistry for this work	[46]

Fig. 7.5 French stamp commemorating the discovery of quinine by Joseph Pelletier and Joseph Caventou, the chemists who first isolated and named chlorophyll

At least two of the researchers on chlorophyll named above have been philatelically commemorated as shown in Fig. 7.5. However, since the inseparable working pair of Pelletier and Caventou made their mark in many other ways, and in particular by discovering quinine in 1820 (2 years after chlorophyll), they have been remembered for that feat by the French Republic.

Today we know there are many types of pigments serving various functional roles in different kinds of photosynthetic organisms. The chlorophylls are designated from a through d, and the bacteriochlorophylls a through g, in order of their discovery. Carotenoids (to be discussed below) and bilins are two other major classifications of photosynthetic pigments. How these pigments work in transforming energy via the photosynthetic process, which may generally be defined as energy capture by an organism for the purpose of driving cellular reactions, spans many disciplines other than chemistry. In examining chlorophyll-based processes, we note that they are light-driven, they utilize the chlorophyll pigments for energy transfer, they are confined to the photosynthetically active radiation region which corresponds exactly to the visible region of the spectrum, 400–700 nm, and they have historically been an active area of research since the sixteenth century. Hence, chlorophyll research rightly belongs to the physicist, biologist, science historian, botanist, plant physiologist, the evolutionary biologist, and so on. However, we must never forget that if chlorophyll did not exhibit a bright green color, it may not have enjoyed center stage in scientific research over the past five centuries.

7.5.2 The Carotenoids and Xanthophylls

The red, orange and yellow pigments that we lump together as the carotenoids and xanthophylls have also piqued the interest of many talented chemists, and can therefore count as many Nobel Prizes as can chlorophyll research as the results of research in this field. There are hundreds of these pigments known today, so we must necessarily confine ourselves to the highlights of this field of research.

Ironically enough, the discovery of carotene is analogous to the synthesis of mauveine (instead of the sought-for quinine) by W. H. Perkin: a German pharmacist named Wackenroder set out to extract an effective anthelminthic from carrots and he found the bright red crystals of a virtually pure pigment instead—a pigment that would later come to be known as "carotene" [47].

Since the pathways by which many of the carotenoids were discovered covers several centuries and is somewhat convoluted, a tabular form for this history is very convenient as well.

Some observations on the last three entries in Table 7.2 are appropriate here. Willstätter seemed to minimize his work on carotenoids in favor of the recognition of his work on chlorophyll, as can be inferred from the little space that he devotes to the former subject in his memoirs [60], and yet his advances in this area opened the pathway to many more Nobel Prizes than those mentioned here, those of Karrer and Kuhn.

Upon presenting the Nobel Prize to Karrer, Wilhelm Palmaer, chair of the prize committee at the time, described him as a scientist with the "ability to visualize great and important problems as well as their smaller parts" and one who in his own unique way "approached problems and pursued new ideas by using his own methods [61]". Karrer's methodology has borne much fruit over the decades. The spinoffs from his work on the carotenoids and xanthophylls is still evolving today: intense research on vision, vitamins, hormones, metabolic pathways, and enzymes.

Much of the same could be said about the work of Kuhn, but there is a darker side to this story. Despite Kuhn's early liberal upbringing, despite the fact that his Ph.D. mentor, Willstätter, was Jewish, and despite the fact that he never actually joined the Nazi party, he nevertheless took it upon himself to align himself with the policies of the Third Reich and to denounce some of his Jewish colleagues even though he was fully aware of the atrocities being committed by this government [62]. Ironically, despite his allegiance to Hitler's policies, Kuhn fell under the same interdict as Domagk and was forbidden to accept his prize until some years after the war was over.

Ideally, Nobel Prizes are awarded to those individuals whose work "during the preceding year, shall have conferred the greatest benefit on mankind," to quote from Alfred Nobel's will. One therefore must wonder why chemical structure determination and chemical syntheses, elegant as they may be in overcoming the problems associated with perhaps a dozen steric centers, are deemed to be of greatest benefit to our human species. It might be difficult to convince the general public of these benefits unless it could be pointed out that many of these esoteric discoveries have opened the door to increasing food production worldwide, to defeating certain molecular diseases, to easing the grip that malaria had on whole populations in many countries, to understanding how vision works and how to correct diseases of the eye, to synthetic production of vitamins to enhance human diets worldwide—all applications that may not have been foreseen at the time of the awarding of the prize! And yet, the Nobel committees continue to view the most fundamental and theoretical breakthroughs in science as potential benefit to all of us, and thus continue to maintain and promote Nobel's far-seeing legacy. Not least among these benefits, as can be seen from much of the preceding chapter, centered around research on colored compounds—the chemical history of color has come a long way indeed!

Table 7.2 Milestones in carotenoid and xanthophyll research

Date	Scientist(s)	Accomplishment	Remarks	References
1831	Wackenroder (1798–1854)	Isolation of crystals of carotene from the juice of pressed out carrots	The pigment, odorless and tasteless, dissolved in ether but not in water	[47, 48]
1847	Zeise (1789–1847)	Zeise found that carotene was soluble in CS_2; the crystals melted at 168 °C.	He recognized that he was dealing with a hydrocarbon but it was impure	[48–50]
1886	Arnaud (1853–1915)	Calculated an empirical formula for carotene, $C_{26}H_{38}$; close to theory; also devised a colorimetric method for determination	Found that carotene crystallized in thin rhombic plates, exhibited dichroism, and was easily oxidized and halogenated	[48, 51, 52]
1886	Lieben (1836–1914)	Identification of carotene in animal tissues; first chemist to study carotenoids in animal tissues	Spectroscopically studied the luteal pigments; called their extract "hemolutein"	[48, 53]
1868–1869	J. L. W. Thudichum (1829–1901)	Isolated yellow crystalline pigment from plants and called it "luteine."	First to define luteine and related pigments as a new class of organic compounds	[48, 54]
1907–1913	Willstätter (1872–1942)	Identified carotene as a hydrocarbon and assigned an empirical formula of $C_{40}H_{56}$	Distinguished carotene from oxygen-containing xanthophyll; 1915 Nobel Prize in Chemistry	[48, 55]
1929	Karrer (1889–1971)	Determined structures of carotene and lycopene; found that they consisted of eight symmetrically arranged isoprene units	Derived Vitamin A from carotene and elucidated its structure, a first for a vitamin; 1937 Nobel Prize in Chemistry	[56, 57]
1930–1940s	Kuhn (1900–1967)	Over a period of 20 years Kuhn discovered eight new types of carotenoids and analyzed their constitution	Kuhn determined the formulas of many vitamins and also synthesized inhibiting factors, or antivitamins. 1938 Nobel Prize in Chemistry	[58, 59]

In the next, and final, chapter, we will see how far the chemical history of color may be able to take us as we look at some modern developments and peer into the future.

References

1. McGrayne SB (2001) Prometheans in the lab: chemistry and the making of the modern world. McGraw-Hill, New York, pp 27–28
2. Chevreul ME (1830) Leçons de chimie appliquée à la teinture, vol 2. Pichon et Didier, Paris, pp 66–76
3. Schweppe H (1997) Indigo and woad. In: Fitzhugh EW (ed) Artists' pigments: a handbook of their history and characteristics. The National Gallery of Art, Washington, DC and Oxford University Press, New York, pp 80–107. As of 2011, 50,000 tons of indigo are produced each year, 95% of which is used to dye the cotton yarn that will eventually become "blue jeans."
4. Fruton JS (1972) Molecules and life: historical essays on the interplay of chemistry and biology. Wiley, New York, p 279
5. Brett RH, Bird G (1835) On the action of acids on the blood productive of certain new substances. Lond Med Gazette 16:751–754
6. Lecanu LR (1838) Études chimiques sur le sang humain. Ann Chim 2e Sér 67:54–70
7. Bernard C (1859) Leçons sur les propriétés physiologiques et les alterations pathologiques des liquids de l'organisme, vol 1. Ballière, Paris, p 254
8. Hoppe-Seyler F (1866) Beiträge zur Kenntnis der Constitution des Blutes. 1. Über die Oxydation im lebenden Blute. Med chem Unt:133–140
9. Stokes G (1819–1903) (1864) On the reduction and oxidation of the colouring matter of the blood. Proc Roy Soc 13:355–364; 357
10. Gerlach J (1858) Mikroskopische Studien aus dem Gebiete der menschlichen Morphologie. Ferdinand Enke, Erlangen, pp 1–4
11. Krafts KP, Hempelmann E, Oleksyn BJ (2011) The color purple: from royalty to laboratory, with apologies to Malachowski. Biotechnic Histochem 86:7–35
12. Fruton JS (1972) Molecules and life: historical essays on the interplay of chemistry and biology. Wiley, New York 195
13. Zacharias E (1881) Über die chemische Beschaffenheit des Zellkerns. Botanische Zeitung 39:169–176
14. Flemming W (1882) Zellsubstanz, Kern und Zelltheilung. F.C.W. Vogel, Leipzig, p 129
15. Kasten FH (1996) Paul Ehrlich: pathfinder in cell biology. 1. Chronicle of his life and accomplishments in immunology, cancer research, and chemotherapy. Biotech Histochem 71(1):2–37
16. Ehrlich P (1878) Contribution to the theory and practice of histological staining. In: Himmelweit H (ed) The collected papers of Paul Ehrlich, 1956–1960. Pergamon Press, London, pp 73–74
17. Parascandola J, Jasensky R (1974) Origins of the receptor theory of drug action. Bulletin Hist Med 48:199–220
18. Marquardt M (1951) Paul Ehrlich. Schuman, New York, p 18
19. Ehrlich P (1885) Sauerstoff-Bedürfnis des Organismus (The need of the organism for oxygen). Habilitationschrift. August Hirschwald, Berlin, p 9
20. Witkop B (1982) Paul Ehrlich: his ideas and his legacy. In: Bernhard CG, Crawford E, Sörbom P (eds) Science, technology and society in the time of Alfred Nobel. Pergamon Press, Oxford
21. Mazumdar P (1974) The antigen-antibody reaction and the physics and chemistry of life. Bulletin Hist Med 48:1–21

22. Langley JN (1901) Observations on the physiological action of extracts of the supra-renal bodies. J Physiol 27:237–256
23. Ehrlich P (1913) Address in pathology on chemotherapy. British Med J 1:354
24. Béchamp A (1863) C R Acad Sci 56:1172–1175
25. Ehrlich P, Bertheim A (1907) Ber Dtsch Chem Ges 40:3292–3297
26. Stadler A-M, Harrowfield J (2011) Places and chemistry: Strasbourg—a chemical crucible seen through historical personalities. Chem Soc Rev 40:2061–2108
27. Lloyd NC, Morgan HW, Nicholson BK, Ronimus RS (2005) The composition of Ehrlich's Salvarsan: resolution of a century-old debate. Angew Chem Int Ed 44:941–944
28. Levinson AS (1977) The structure of Salvarsan and the arsenic-arsenic double bond. J Chem Educ 54:98–99
29. Travis AS (1989) Science as receptor of technology: Paul Ehrlich and the synthetic dyestuffs industry. Sci Context 3:383–408
30. Paraphrased from a letter of Paul Ehrlich to Christian Herter, and quoted by the latter in "Imagination and Idealism in the Medical Sciences," an address given at the opening of the Columbia University Medical School, 23 September 1909; as given by Dale HH (1956) Introduction. In: Himmelweit F, Marquardt M, Dale H (eds) The collected papers of Paul Ehrlich. Histology, biochemistry and pathology, vol I. Pergamon Press, London, pp 1–18; 9
31. Bäumler C (1984) Paul Ehrlich: Scientist for Life. Holmes & Meier, New York; p 62
32. Domagk G—biography. Nobelprize.org. 14 Dec 2011. http://www.nobelprize.org/nobel_prizes/medicine/laureates/1939/domagk-bio.html. Accessed 14 Dec 2011
33. Wood ME (2010) Milestones: soldier sulfa. Chem Heritage 28(1):7
34. Müller P—biography. Nobelprize.org. 14 Dec 2011. http://www.nobelprize.org/nobel_prizes/medicine/laureates/1948/muller-bio.html. Accessed 14 Dec 2011
35. Sadasivaiah S, Tozan Y, Breman JG (2007) Dichlorodiphenyltrichloroethane (DDT) for indoor residual spraying in Africa: how can it be used for malaria control? Am J Trop Med Hyg 77(Suppl 6):249–263
36. Cambrosio A, Keating P (1992) Between fact and technique: the beginnings of hybridoma technology. J Hist Biol 25:175–230
37. Köhler G, Milstein C (1975) Continuous cultures of fused cells secreting antibody of predefined specificity. Nature 256:495–497. It is interesting to note that the editors of *Nature* saw nothing exceptional in this paper. Although it was submitted as a regular article, it was relegated to the "Letters to *Nature*" section, and preceded by 15 other letters on such earth-shaking subjects as visual functions in goldfish and defensive stoning by baboons
38. Jerne NK—autobiography. Nobelprize.org. 14 Dec 2011. http://www.nobelprize.org/nobel_prizes/medicine/laureates/1984/jerne-autobio.html. Accessed 14 Dec 2011
39. Waldmann TA (1991) Monoclonal antibodies in diagnosis and therapy. Science 252: 1657–1662; 1657 (reprinted with permission from AAAS)
40. Blankenship RE (2002) Molecular mechanisms of photosynthesis. Blackwells Scientific Ltd, Oxford
41. Délepine M (1951) Joseph Pelletier and Joseph Caventou. J Chem Educ 28:454–461
42. Krasnovsky AA (2002) Chlorophyll isolation, structure and function: major landmarks of the early history of research in the Russian Empire and the Soviet Union. Photosynth Res 76:389–403
43. Willstätter RM (1920) On plant pigments. Nobel lectures. Elsevier, Amsterdam, pp 301–312
44. Tswett MS (1906) Physikalish-chemische Studien über das Chlorophyll. Die Absortionnen. Ber Deutch Bot Ges 24:316–323
45. Fischer H—biography. Nobelprize.org. 15 Dec 2011. http://www.nobelprize.org/nobel_prizes/chemistry/laureates/1930/fischer-bio.html. Accessed 15 Dec 2011
46. Woodward RB (1972) Recent advances in the chemistry of natural products. Nobel lectures, Chemistry 1963–1970. Elsevier, Amsterdam, pp 100–121
47. Wackenroder H (1831) Ueber das Oleum radicis Dauci aetherum, das Carotin, den Carotenzucker und den officinellen succus Dauci; so wie auch über das Mannit, welches in

dem Möhrensafte durch eine besondere Art der Gährung gebildet wird. Geigers Magazin der Pharmazie 33:144–172

48. Sourkes TL (2009) The discovery and early history of carotene. Bull Hist Chem 34(1):32–38 (This paper is cited in Table 9.2 along with the original sources which are contained in the "References and Notes")

49. Zeise WC (1847) Ueber das Carotin. Ann Chem Pharm 62:380–382

50. Zeise WC (1847) Einige Bemerkungen über das Carotin. J Prakt Chem 40:297–299

51. Arnaud A (1886) Recherches sur la composition de la carotine, sa fonction chimique et sa formule. C R Séances Acad Sci Ser C 102:1119–1122

52. Arnaud A (1889) Recherches sur la carotine, son role physiologique probable dans la feuille. C R Séances Acad Sci Ser C 109:911–914

53. Piccolo G, Lieben A (1886) Studi sul corpo lutea della vacca. Giornale di Scienze Naturali ed Economiche (Palermo) 2:25

54. Drabkin DL (1958) Thudichum, Chemist of the Brain. University of Pennsylvania Press, Philadelphia

55. Willstätter RM, Mieg W (1907) Ueber die Gelben Begleiter des Chlorophylls. Justus Liebigs Ann Chem 355:1–28

56. Karrer P (1966) Carotenoids, Flavins and Vitamin A and B_2. Nobel lectures: chemistry, 1922–1941. Elsevier, Amsterdam, pp 443–448

57. Miller JA (1993) 1937 Nobel Laureate: Paul Karrer (1889–1971). In: Miles W (ed) Nobel Laureates in chemistry, 1901–1992. American Chemical Society, Washington, DC, pp 242–246

58. Kuhn R—biography. Nobelprize.org. 18 Dec 2011. http://www.nobelprize.org/nobel_prizes/chemistry/laureates/1938/kuhn-bio.html. Accessed 18 Dec 2011

59. Miller JA (1993) 1938 Nobel Laureate: Richard Kuhn (1900–1967). In: Miles W (ed) Nobel Laureates in chemistry, 1901–1992. American Chemical Society, Washington, DC, pp 248–252

60. Willstätter RM (1965) From my life. E.A. Benjamin, New York (Translated from the German by LS Hornig)

61. Karrer P (1966) Carotenoids, flavins and vitamin A and B2. Nobel lectures: chemistry, 1922-1941. Elsevier, Amsterdam, pp 443–448, 407–413

62. Deichmann U (2001) German-Jewish chemists and biochemists in exile. In: Szöllösi-Janze M (ed) Science in the third Reich. Oxford International Publishers, Oxford, pp 264–265

Chapter 8
Finale: Color in Foods, Photochemistry, Photoluminescence, Pharmaceuticals, Fireworks, Fun, and the Future

This final chapter will alliteratively pick up many topics that fell outside the trajectory traced by the history–chemistry—color interface in the previous seven chapters. We will see how colored additives affected the food quality of the past and, by extension, how color has affected, and continues to influence, so many other aspects of our daily lives.

8.1 Color in Foods

When creatures in the wild see a brightly colored snake or frog, they realize immediately that the colors are a signal—"Don't come near me! Don't try to take a bite out of me! I am deadly poisonous to you." One can say the same thing about most brightly colored mineral pigments: attractive to the eye they certainly are, looking good enough to eat—but deadly to the one who dines on them.

In 1820 and later in the 1850s in Britain, Accum (1769–1838) and Hassall (1817–1894), respectively, blew the whistle on the practice of food adulteration, much to the chagrin of the food purveyors of the day. Much of the practice had the goal of adding bulk and weight to the food, such as bread, with less expensive, and often non-nutritious or harmful, additives. But one goal for many producers was to make the food attractive—a practice carried out today in a much more sophisticated manner.

Coloring food and drink in order to deceive is certainly not a new practice. Pliny the Elder in the third century BCE comments on the wine industry in Roman Gaul [1]:

> …about the rest of the wines grown in the province of Narbonne no positive statement can be made, in as much as the dealers have set up a regular factory for the purpose and color them by means of smoke…a dealer actually uses aloe for adulterating the flavor and color of his wines.

M. V. Orna, *The Chemical History of Color*, SpringerBriefs in History of Chemistry, DOI: 10.1007/978-3-642-32642-4_8, © The Author(s) 2013

Combatting such practices was a long time in coming. The first recorded "pure food laws" were enacted in thirteenth century Europe [2], but enforcement always lagged behind the actual law, and things really got out of hand with the great trade expansion of the sixteenth and seventeenth centuries when all kinds of exotic and expensive brews and spices were imported into Europe.

Toward the end of the eighteenth century, Accum, a German chemist residing in England, began to publish a series of articles on food adulteration in *Nicholson's Journal* which culminated in the publication of his 1820 comprehensive treatise on this subject [3]. In it, he documented how certain foods contained highly toxic adulterants for the purpose of making them attractive, especially to children. Some of the favorite colored adulterants were red lead, vermilion (HgS), and copper salts (to make pickles a luscious bright green!). Accum's chapter on Poisonous Confectionery opens thus [4]:

> *In the preparation of sugar plums, comfits, and other kinds of confectionery, especially those sweetmeats of inferior quality, frequently exposed to sale in the open streets, for the allurement of children, the grossest abuses are committed... the red sugar drops are usually coloured with the inferior kind of vermilion. The pigment is generally adulterated with red lead. Other kinds of sweetmeats are sometimes rendered poisonous by being coloured with preparations of copper.*

He closes this chapter, as he does many others, by indicating how one can detect the possible adulterant by physical or chemical means.

Later in the century, Hassall's assiduous work [5], which included photomicrographs of plant parts in order to illustrate some types of organic additives, documented adulterants of virtually every kind in all types of food and drink. He admitted that one would expect to find copper salts in many preserves, but went on to say [6]:

> *But the still larger quantities of copper detected in certain of the samples of greengage jam seemed to show that...some greening salt of copper...is really intentionally introduced for the purpose of creating an artificial viridity...Although we may fairly expect to find copper in any preserved vegetable substance prepared in the ordinary manner, yet we scarcely expected to meet with that poison in those tasteful and sparkling little boxes of bonbons which at Christmas-time are displayed in shop windows so attractively.*

This early work by two nineteenth century pioneers resulted in the passage of stringent food laws applied to both organic and inorganic additives. Nonetheless, toxic substances such as copper(II) sulfate and lead chromate could be found in foods well into the twentieth century.

Furthermore, after the Perkin "revolution," the practice adding adulterants and colorants to food changed very rapidly with the synthesis of hundreds of bright new dyes. The great concern was the safety of these new dyes for human consumption. It was Harvey W. Wiley (1844–1930), the chief chemist at what is now the U.S. Department of Agriculture, who was the primary author of the Pure Food and Drugs Act of 1906, an act that recognized seven coal tar dyes as safe for use in foods (see Fig. 8.1). Five of those dyes have since been de-listed, and the survivors, erythrosine and indigotine, have been joined by six others in compliance

Fig. 8.1 The 3-cent postage stamp issued by the U.S. Post Office in Wiley's honor on June 27, 1956, in conjunction with the fiftieth anniversary of the 1906 Act

with the Federal Food, Drug and Cosmetic Act of 1938 [7]. Present food additive practice is summarized in the U.S. FDA document "Food Ingredients and Colors" [8].

Despite such controls, there is mounting public concern that today's practices of artificially coloring meats, fruits, and vegetables can lead to behavioral disorders in children and possibly a whole host of other disorders in adults. The whole foods/ organic foods movement has its basis in this fear. As a result, there is a growing tendency to "go natural" with respect to added colorants such as using anthocyanins, compounds that in addition to their rich, red hue can provide an amazing assortment of health benefits: pain relief, antioxidant activity, and inhibition of inflammation [9, 10]. Another natural colorant, saffron, is often used to impart a rich yellow color and distinct flavor to traditional Spanish paella, French bouillabaisse, and other dishes. However, saffron is known as the most expensive spice in the world: it is harvested just three delicate stigmas at a time from the purple flower of the saffron crocus (*Crocus sativus*). Producing just one pound of saffron requires hundreds of thousands of stigmas, thousands of flowers, and hundreds of hours of meticulous labor to garner the yellow-producing carotenoid, crocin, and other coloring and flavoring substances [11]. So going natural can have its down side as well. But there is no question that color in foods has been a major concern from prehistoric times down to our own day.

8.2 Photochemistry and Photoluminescence

There are many processes in nature that are essential to our lives that involve photochemistry. The oxygen we breathe is produced by photosynthesis. Our entire food chain depends on photosynthesis. The protective ozone layer surrounding the earth depends on photochemical synthesis. Photochemistry in the retina of the eye allows us to see. The warmth of the earth's surface depends on photochemical absorption of light that is then dissipated as heat. The section that follows illustrates some photochemical phenomena that have a history of deliberate human intervention, leading to theoretical and practical discoveries.

Emission of absorbed light as radiation leads to the phenomenon of photoluminescence, which is important in many areas of science and technology. It will be treated in Sect. 8.2.2.

8.2.1 Film and Photography

The need to record events and persons as well as abstract concepts is one that is very evident in the history of art. Between the earliest cave painter and the most sophisticated portrait painter of the nineteenth century stands a long tradition of recording almost any subject matter conceivable. It is only natural that inventors should experiment with the means of producing highly realistic images and should devise ways of preserving them in permanent records. It was not long before the combination of these two processes in a new representational tool, photography, was developed and adopted by artists as an aid to art. Eventually, it grew into an art form in its own right.

No technical process evolves by itself. Evolution in auxiliary, and sometimes in seemingly unrelated areas, is necessary for the final coming together of technical achievement. Photography was no exception to this. Although some of the basic principles of optics were known since the time of Aristotle, the elements of image production began to coalesce in the sixteenth century with the invention of the *camera obscura*, lenses, the microscope and the telescope. Thus the essential ingredients of the modern photographic camera and the principles of image formation were well understood over 400 years ago, but it took another 300 years before anyone succeeded in making a permanent record of these optical images. This could only happen with the evolution of chemical science.

In order to render an optical image permanent, it is necessary that the light forming the image also have the power to produce a chemical change in a light-sensitive material medium. Silver compounds have been known since at least the thirteenth century, and it has long been known that the halides of silver, $AgCl$, $AgBr$ and AgI, are very sensitive to light. Around the beginning of the nineteenth century, Thomas Wedgwood (1771–1805), of the porcelain-manufacturing family, and Sir Humphry Davy (1778–1829), were among the first to produce images on paper impregnated with silver nitrate or silver chloride, but neither discovered a way of preventing their papers from darkening upon exposure to light. Progress in this area was made by Nicephore Niepce (1765–1833) and Louis J. M. Daguerre (1787–1851), but the first person to actively use silver bromide was John F. Goddard (1795–1866) in 1840, thus greatly improving on Daguerre's method, allowing photographic exposure time to be reduced to just a few seconds. Meanwhile, a physician named Richard Leach Maddox (1816–1902) developed

Fig. 8.2 Absorbance curves of film sensitizing dyes. *1* Black and white film (unsensitized); *2* Blue-sensitive dye; *3* Green-sensitive dye; *4* Red-sensitive dye. © 1998, M. V. Orna

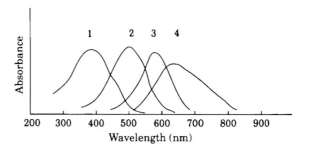

the photographic emulsion, gelatin impregnated with silver halide which, with some refinements, remains essentially the same to this day [12, 13]. The great breakthrough came when in 1841 William Henry Fox Talbot (1800–1877) discovered the latent image, an image already present, though invisible, on photographic emulsions exposed to a relatively weak light signal but amplified enormously by "development" by a mild reducing agent. Though Fox Talbot receives the credit for being the "Father of Modern Photography," John Herschel (1792–1871) who, 37 years before had discovered the superiority of silver bromide to silver nitrate [14], was working on the problem simultaneously and discovered the ability of sodium thiosulfate to "fix" the photograph to prevent further darkening [15]. Herschel never sought monetary gain for his inventions and he generously passed his knowledge on to Fox Talbot who had no compunctions about including it in his patents [14].

While black and white photography was well under way using the processes described above, recording images in color remained a vexing problem. A first feeble step was made by Carl Wilhelm Scheele (1742–1786) when in 1777 he discovered that the various colors of visible light were not equally effective in darkening silver halides, blue light being far more effective than the longer wavelengths of light [16]. The next step in the development of color photography came in the field of color-vision theory, principally through the work of Thomas Young (1773–1829) in 1802 (Sect. 2.6.4). Then, in 1861, James Clerk Maxwell (1831–1879) was the first person to achieve a color image, but it was fraught with inconveniences and shortcomings. Perhaps the greatest of these was the fact that silver halides have the maximum sensitivity to ultraviolet and short wavelength visible light, whereas the longer wavelengths of visible light, in the red region of the spectrum, hardly affect them. Something more was needed to enhance the sensitivity in the red and green regions of the spectrum.

In 1873, Hermann Wilhelm Vogel (1834–1898) discovered that a yellow dye could make a silver halide photographic emulsion more sensitive to the green region of the spectrum, but it was not until a generation later, in 1904, that the production of "panchromatic" film was made possible through fresh dye discoveries. The breakthrough dye, pinacyanol, was discovered in 1904 by Benno Homolka (1860–1925) of the Hoechst Dye Works, and this dye remained the most important sensitizer for the entire range of the visible spectrum until well into the 1930s, basically because progress in this area was very slow [17]. Figure 8.2

Table 8.1 Structures of some typical sensitizing dyes

Number	Color of dye	Color sensitivity	Name of dye
0	Yellow	Blue	Cyanine
1	Magenta	Green	Carbocyanine
2	Cyan	Red	Dicarbocyanine

depicts the light sensitivity curve of ordinary silver halide and the sensitivity that can be achieved for color photography by addition of blue-sensitive, green-sensitive and red-sensitive dyes to the film emulsion. Since the range of maximum absorbance (between 300 and 400 nm for silver halides) indicates the range of maximum sensitivity to light, silver halides will tend to absorb light mainly in the 300–400 nm range. The blue-, green- and red-sensitive dyes have absorption maxima at longer wavelengths. Dyes have also been developed that can extend the range of sensitivity of film to around 1,300 nm, well into the infrared region of the spectrum. Table 8.1 denotes the structural formulas for some representative sensitizing dyes.

A general theory of silver halide sensitization involves the idea that the sensitizing dye acts as an intermediate between the impinging photons of light and the silver halide grains in the formation of the latent image. Once the sensitization problem was overcome, the way lay open for the manufacture of color photographic film, and of color processing to form a dye image, rather than a silver image, in the final product, processes well beyond the scope of this book. References [18–20] can beautifully complete the fascinating story of color photographic processing.

With the advent of digital photography, photographic film now occupies a niche market, and photography, once the realm of the chemist, has returned to the open arms of the physicist. The digital camera, in a highly sophisticated way, re-enacts the Maxwell original color photographic process by using a set of three primary color filters (red, green, and blue) to record the light reflected from the subject and transform it into a digital representation that can be transmitted electronically. "How Stuff Works" is a good introductory tutorial on this subject [21].

8.2.2 Photoluminescence: Fluorescence and Phosphorescence

Qualitative aspects of photochemistry were known long before the development of chemistry as a science. For example, the history of chemical luminescence (or chemiluminescence) goes back to 1603 when Vincenzo Cascariolo, an alchemist

Fig. 8.3 *E* Excitation,
F Fluorescence (allowed,
fast), *U* Unfavorable (slow),
P Phosphorescence
(forbidden, slow).
© 2010, M. V. Orna

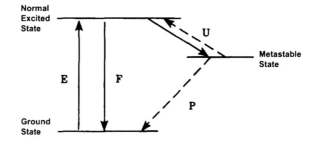

in Bologna, Italy, heated barite (barium sulfate) and coal together and observed that upon cooling the powder exhibited a bluish glow in the dark and that the glow could be reactivated by exposure of the powder to sunlight. He apparently had made barium sulfide, a well-known phosphorescent material, though his powder became known as *lapis solaris* (sunstone) or *phosphorus* (light bearer). Now we reserve the name phosphorus for the element, which does exhibit phosphorescence, and call microcrystalline luminescent materials phosphors [22].

To place in context this first recorded discovery of a phosphor, we must recall (from Sect. 6.3) that Gustav Kirchhoff in 1859 realized that the observed frequencies of the various elements' emission lines in their bright line spectra corresponded to the frequencies observed in Fraunhofer's dark line spectra. Kirchhoff concluded that the dark lines were due to the absorption of the characteristic frequencies of the elements present in the cooler outer layers of the sun's atmosphere, and that these were the same frequencies that these elements emitted when excited by an energy source such as a flame. In other words, he observed the resonance lines, radiation absorbed by a substance and immediately emitted at the same wavelength.

Other substances, such as barium sulfide, can absorb a certain wavelength of light, but because of the molecular energy level scheme and the quantum mechanical selection rules which govern energy level transitions, the absorbed radiation may be emitted at a longer wavelength and there may also be a time delay in the transition. An examination of Fig. 8.3 will help clarify this statement and also depict the difference between phosphorescent and fluorescent materials.

When radiation is absorbed by excitation (E), a number of pathways exist for the removal of the energy of the excited states, most often by radiationless transitions; therefore, most materials do not fluoresce or phosphoresce. Normal excited states that almost immediately revert back to the ground state, often spontaneously emitting light of lower frequency depending upon the fine structure of the excited state, are said to exhibit fluorescence (F). Other excited states, which are metastable, also exist. If after excitation, energy transfer to such a metastable state occurs (U), relaxation back to the ground state is very slow (P). "Allowed" and "forbidden" are quantum mechanical terms that give an indication of the probability of a reaction; a "forbidden" process is a highly improbable process. A "back of the envelope" way of distinguishing between a fluorescent and a phosphorescent substance is to shine a light on the substance and then turn it off. If the substance glows (emits radiation) only while the light is turned on, it is fluorescent; if there is a time delay and the glow persists after the source of excitation is removed, it is phosphorescent. Both substances are said to be photoluminescent, that is, luminous by virtue of the source of excitation, photons. (However, there is an overlap in the lifetimes of the excited states of both fluorescent and phosphorescent materials, so to be perfectly precise, a phosphorescent substance must have passed through an intermediate state upon relaxation.) Other forms of luminescence, distinguished by the mode of excitation, are chemiluminescence, bioluminescence, electroluminescence, triboluminescence, etc.

Observation of luminescent phenomena has a long history punctuated by the work of famous individuals like Hennig Brand (c. 1610–c. 1730), who discovered phosphorus in 1677, and Haüy (1743–1822), the French mineralogist who observed the double colors of some crystals of fluorite [23]. Then in 1852, at almost the same time that Kirchhoff observed resonance emission of radiation, Stokes (1819–1903) published his famous paper [24] that led to the formulation of Stokes' Shift and eventually to his adoption of the term fluorescence [25] for the phenomenon that the emitted light has a longer wavelength, and therefore lower energy, than the incident radiation in a photoluminescent event. One of the earliest applications of fluorescence was that of Alexandre-Edmond Becquerel (1820–1891) in 1857, who had the idea of coating the inside of an electric discharge tube with a fluorescent material [26], thus anticipating the development of the mercury vapor lamp in the early twentieth century by Peter Cooper Hewitt (1861–1921) and subsequent development of modern fluorescent lighting. Please see the excellent review article on this subject in the *Journal of Chemical Education* [27].

8.2.3 Photoluminescence: Lasers

In his 1917 paper, *Zur Quantentheorie der Strahlung* [28], Albert Einstein developed the theoretical basis of the absorption, spontaneous emission, and also the stimulated emission of electromagnetic radiation. These processes depend upon the probability coefficients, sometimes called the "Einstein coefficients," that govern these so-called "allowed" and "forbidden" transitions. We have seen that light absorption is highly probable (allowed) as is fluorescence, although the spontaneously emitted radiation is non-coherent, i.e., emitted in all directions. If, however, the emission can be stimulated by quanta of light of the exact same wavelength as the emitted radiation before spontaneous emission occurs, then the stimulated emission can be induced at once, and if it takes place in a suitably constructed cavity where the emissions can multiply with a type of "chain reaction" effect, then we will have constructed a device capable of Light Amplification by Stimulated Emission of Radiation, in other words, a LASER.

The first "laser" did not actually use visible light, but microwave radiation, and hence it was termed a "maser." This device was developed by Charles H. Townes (b. 1915) in 1953, and in 1964 he shared the Nobel Prize in Physics with Nikolay

Basov (1922–2001) and Aleksandr Prokhorov (1916–2002), two Russian physicists who independently discovered how to produce continuous output, something that Townes was unable to do. The three were cited "for fundamental work in the field of quantum electronics, which has led to the construction of oscillators and amplifiers based on the maser-laser principle" [29]. Since that time, many different types of lasers were developed—gas, solid-state, fiber, semiconductor, dye, etc.— and they have found use in hundreds of applications in virtually every field of endeavor such as the military, industry, law enforcement, medicine, entertainment, and basic research [29].

Chemistry enters the picture with the advent of the tunable dye laser, a device with a variable wavelength using solutions of organic dyes as the lasing medium. Some typical laser dyes are Oxazine 9, Rhodamine B, Coumarin 9, Stilbene, Fluorescein, and Malachite Green, among others [30]. They are especially useful in medicine, particularly dermatology, because their tunability allows them to closely match the absorption characteristics of certain tissues such as hemoglobin [31].

8.2.4 Color in Fireworks: Pyrotechnics

Green, red and blue fireworks owe their coloring to the presence of barium, strontium, and copper salts respectively. Originating in China in the seventh century, the first fireworks are believed to have consisted of KNO_3 (saltpeter), carbon and sulfur packed into hollowed bamboo shoots which created a loud explosion when ignited. In the eighteenth century, a Jesuit missionary to China, Pierre Nicolas le Chéron d'Incarville, described how Chinese fireworks were made to the Paris Academy of Sciences [32]. Publication of the book served to popularize the use of fireworks for celebrations, a practice that later became a "must" in the July 4 festivities in the United States.

In modern fireworks a combination of heavy metals and perchlorate oxidizers delivers the color and intensity that are widely synonymous with the start of a new year. However, recent environmental studies suggest that the formulation of most fireworks contributes to soil and watershed contamination. Chemists have developed nitrogen-rich, environmentally benign pyrotechnics that, upon detonation, mimic the color intensity of perchlorate-based fireworks while minimizing their negative impact on the surrounding ecosystems [33]. New pyrotechnic compounds and formulations such as derivatives of the tetrazoles and tetrazines are about to revolutionize traditional pyrotechnic compositions [34]. New research has also shown that copper compounds can substitute for noxious barium compounds to produce green colored effects, but the application is limited because of copper's noble and often hygroscopic character of its compounds. Steinhauser and Tarantik investigated nitrocuprates(II) as well as basic copper(II) nitrate and found that a formulation based on ammonium nitratocuprate(II) nitrate and boron looks promising [35].

8.3 Color in Pharmaceuticals: Pills and Tablets

"Good morning dear unseen audience everywhere! This is Vera Cheera coming to you with her morning sunshine talk through the courtesy of Plunkett's Pink Pills for Pale People..." Thus began the 1930s radio show, "Vera Cheera's Morning Sunshine Talk," a running ad for the pink pills useful for all things cosmetic and healthful for the modern woman [36].

Lest you think pink pills were around for a long time, think again. Medicinal concoctions, usually rolled into balls and mixed with clay or with bread as fillers, were delivered as spherical "pills" or "tablets" up until the twentieth century and contained no colored additives. There is evidence that the color of ancient medications was natural, bearing the color of the ingredients, mainly plant parts [37]. There is other evidence that pills could be delivered literally rolled in gold or silver foil as an edible coating (with pricing incremented accordingly) [38]. Only in mid-twentieth century did coloration in medicine come into its own when the pharmaceutical companies discovered the greater marketability of colored products and the advantage of branding. Today Big Pharma has more than 80,000 color combinations to choose from, no longer limited to white or to pink!

8.4 Color for Fun and Enjoyment

Who among us does not enjoy the vision of a rainbow? The spectral colors spread out across the sky inspire awe and pleasure no matter how many times we see them. For this reason, we have attempted to replicate rainbows artificially. We can go to any toyshop and find rainbow glasses, any rock shop and find rainbow quartz made by plasma ionization techniques. Jewelry and craft shops sell rainbow jewelry made by electroplating titanium or niobium oxide. Teachers use rainbow tubes with a universal indicator to demonstrate pH changes and acid–base reactions. Dye mixtures can be separated by rainbow electrophoresis and rainbow chromatography. Rainbow thermometers and toys are made from liquid crystals. Thermochromic materials can be heated to produce rainbow colors. Every toy shop is awash with rainbow colors, a phenomenon not available prior to the many inventions of the twentieth century [39].

It is not clear from the archaeological record if such colors were available for fun and pleasure in ancient times. The oldest known toy was a doll, with the yo–yo coming in a close second, but the coloring matter applied to these artifacts has not survived. What we do know from gravesites is that colored precious and semi-precious stones were an important part of human life and over the centuries, the joy of the color they impart has not abated. We see this joy in the face of a person enthralled by a beautiful amethyst, emerald, or diamond.

As the prerogative of royalty, purple has created a mystique for amethyst that elevated it above other types of quartz crystals. Coveted by Catherine the Great and embedded in the British Crown Jewels, amethyst was once considered a

Fig. 8.4 A sample of
Alaskan amethyst.
Photograph by M. V. Orna

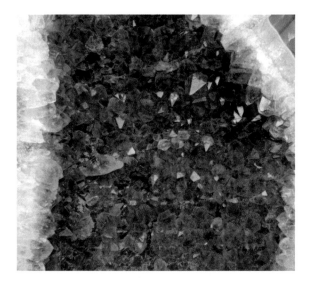

semiprecious gem because of its rarity. Today, increased availability and afford-
able prices have caused it to enjoy an even greater popularity [40]. Figure 8.4 is
fine example of amethyst.

Toward the end of his life, Berthelot (1827–1907) was able to show that
amethyst, on heating, became colorless, but that the color could be restored by
exposition to radium radiation [41]. Crystallographers today have discovered that
"color centers," structural flaws in the crystal lattice, are responsible for trapping
or releasing energized electrons which can change the valencies of the interstitial
iron impurities, thus yielding the colors.

Emerald is another gemstone treasured for its color over the past 4,000 years. It
was reputed to be Cleopatra's favorite gem and even Pliny could wax poetic over
its color. Its name is derived from the Greek *smaragdos*, which simply means
"green gemstone." Chemists have shown that emerald is nothing more than
colorless beryl, aluminum beryllium silicate, with chromium impurities replacing
the aluminum in the crystal lattice. The most highly prized emeralds are those
from the Andes near Bogotá, Colombia. Modern mineralogists have presented
evidence that hydrothermal brines transported sulfates to structurally favorable
sites where it was thermochemically reduced, and that the sulfur generated thus
reacted with organic matter to release trapped chromium, vanadium, and beryllium, which in turn enabled emerald formation [42].

What is the most famous diamond in the world? What is the most visited piece
of artwork apart from the Mona Lisa? To what diamond was attributed the curse
that brought about the executions of Louis XVI and Marie Antoinette? What
diamond appreciated over 100 times its original valuation during the course of the
twentieth century? What diamond was donated by Harry Winston to the Smithsonian National Museum of Natural History? What diamond contains boron
impurities within its carbon lattice that confers on it a deep blue color, its

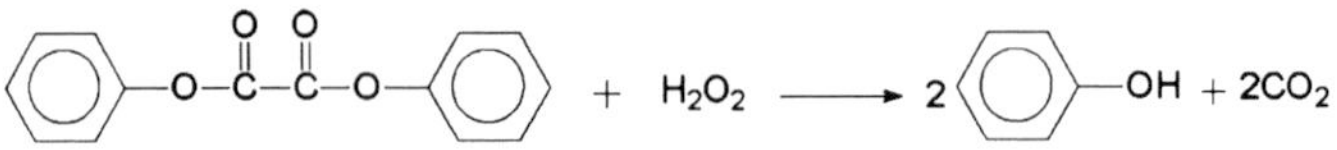

Fig. 8.5 Reaction of phenyl oxalate ester with hydrogen peroxide to produce phenol and carbon dioxide

phosphorescence, and its semiconductivity? What diamond is unique in all the world because it glows an eerie reddish-orange on exposure to ultraviolet light [43]? Of course, the answer to all these questions is "the Hope diamond." Weighing in at 45.52 carats, it is the largest and most perfect blue diamond in existence, older than a billion years, and originally the size of a golf ball [44]. The Hope diamond never ceases to fascinate—and especially chemists. They have shown that at a level of only one or a few boron atoms for every million carbon atoms, an absorption spectrum results that produces an attractive blue color, a very rare and highly prized phenomenon. In addition, since the boron acceptor energy is so small, thermal excitation can take place at room temperature, rendering blue diamonds conductors of electricity as well [45, 46]. The Smithsonian Institution website [47] has a brief history of the Hope diamond with a number of up-to-date references.

Besides toys and gemstones, color gives pleasure in other sectors of nature and the artifices derived from it. Robert Boyle (1627–1691) is given credit for first having scientifically investigated bioluminescence in 1672 when he established that the luminescence emissions from rotting wood, rotting flesh, and fireflies require air, are cold-light systems, and can be inhibited by chemical reagents such as alcohols, ammonia and hydrochloric acid. Since that time, chemists have found that the phenomenon is catalyzed by enzymes. A whole host of other creatures exhibit it, such as glowworms, earthworms, snails, jellyfish, angler fishes, and luminescent fungi. Bioluminescence gives delight to tourists and to deep sea divers as well.

It is not necessary to dust off your scuba-diving equipment to observe luminescent phenomena. Readily available lightsticks [39], marketed under names as "Cyalume," "Glow Light," etc., can produce a chemiluminescent kick by activating two internal substances to produce the glow. The reaction is based on energy transfer between an intermediate formed in the reaction of phenyl oxalate ester with hydrogen peroxide as shown in Fig. 8.5, followed by reaction of the intermediate, thought to be C_2O_4, an unstable molecule, with a typical anthracene-type dye molecule, such as 9,10-bis(phenylethynyl)anthracene, as follows:

$$Dye + C_2O_4 \rightarrow Dye^* + 2CO_2$$

where Dye* represents the activated dye molecule prior to the emission event, which can be written as:

$$Dye^* \rightarrow Dye + h\nu$$

where $h\nu$ represents the emitted photon of light of frequency ν. Further details of this reaction can be found in [48]. Other chemiluminescent reactions have been utilized in a variety of settings, one of the best-known being luminol (5-Amino-2,3-dihydro-1,4-phthalazinedione). When luminol is reacted with an oxidant such as hydrogen peroxide in the presence of a catalyst, it luminesces very strongly. Iron is an ideal catalyst, and since iron is contained in blood, forensic scientists use luminol as a sensitive test for the presence of blood at crime scenes. Other reactions of this type have been published in scientific journals, but almost all of them require not readily available or toxic chemicals. An exception that can be easily done by anyone who wants to have chemiluminescent fun, even in a disco [49], is examination of the quinine in tonic water. Quinine is a fluorescent molecule that emits a blue glow when irradiated with a near-ultraviolet light source such as the black light that strobes around nightlife venues. Addition of salt quenches the fluorescence, and acids such as lemon juice affect the fluorescence differently at different concentrations. So go to your favorite disco, order up some tonic water and a lemon sour, and amaze your friends with a chemical demonstration in a most unlikely place. Chemistry is everywhere, color is everywhere, and having chemical fun with color can also be everywhere!

8.5 Color in the Future

At the 5:00 P.M. rush hour on Friday, December 15, 1967, the 2235-foot suspension bridge spanning the Ohio River between Gallipolis, Ohio and Point Pleasant, West Virginia suddenly collapsed—spilling cars and trucks into the cold December water. Investigations of the Silver Bridge accident revealed the cause to be one corroded pin that had sheared off—a defect, all agreed, not easily detected. But, what if…a paint could be developed that would signal such underlying corrosion by changing color or fluorescing? The paint would need to be sensitive enough to react to the pH increase associated with corrosion yet expansive enough so that the color change could be seen by the naked eye. What if….? [50] Help may very well be on the way through advances in materials science and corrosion engineering—a new role for color and color changes into the future [51].

Quantum dots, or Qdots, seem to be all the rage in the recent literature. They are fluorescent semiconductor nanocrystals with energy band gaps governed by the infinitely deep square-well potential particle-in-a-box model. As the size of the quantum dot decreases, the energy band gap increases, which allows for fine tuning of the wavelengths of light absorbed or emitted, leading to an infinite variety of possible colors. But these colors are not all for fun and games. They have almost as many applications as the variety of their colors, with properties that make them useful as diagnostic tools and monitors for biological processes, including cancer growth. They could also be used for producing inexpensive light-emitting diodes (LEDs) and high efficiency solar cells. Since the dots can be injected into almost any kind of matrix, and since their absorption and emission characteristics can be specifically controlled, they have the possibility of serving as unique validation signatures and identifiers [52, 53].

With the rapidly expanding market in such devices as electronic readers, inexpensive non-light-emitting materials for display applications are in high demand, able to switch from a reflective color to an essentially colorless transmissive state—a property not present in multicolored electrochromic polymers. After 10 years of systematic structure–property development, chemists at the University of Florida have finally made a breakthrough. They are now able to manipulate the composition of thiophene-based polymers to create the first set of soluble materials that exhibit the whole gamut of colors possible. These polymers have great potential for use in virtually every display application, and the commercial applications are endless [54]. And so are the possibilities for color— but since this book will not be endless, the reader is invited to continue observing future developments as they unfold in other sources.

Ben Franklin is reputed to have remarked: "About light, I am in the dark." It is to be hoped that readers of this book can no longer say the same thing about "The Chemical History of Color."

References

1. Rackham H (ed) (1945) Pliny the Elder, Natural History, Col. 4, Book XIV. Harvard University Press, Cambridge, pp 232–233
2. McKone HT (1990) Copper in the candy, chromium in the custard: the history of food colorants before aniline dyes. Today's Chemist 3:22–25, 34
3. Accum F (1820) A treatise on adulterations of food, and culinary poisons exhibiting the fraudulent sophistications of bread, beer, wine, spiritous liquors, tea, coffee, cream, confectionery, vinegar, mustard, pepper, cheese, olive oil, pickles, and other articles employed in domestic economy. Abraham Small, Philadelphia
4. Accum F (1820) A treatise on adulterations of food, and culinary poisons exhibiting the fraudulent sophistications of bread, beer, wine, spiritous liquors, tea, coffee, cream, confectionery, vinegar, mustard, pepper, cheese, olive oil, pickles, and other articles employed in domestic economy. Abraham Small, Philadelphia, p 224
5. Hassall AH (1876) Food: its adulterations and the methods for their detection. Longmans, Green, and Co., London
6. Hassall AH (1876) Food: its adulterations and the methods for their detection. Longmans, Green, and Co., London, p 504
7. Sharma V, McKone HT, Markow PG (2011) A global perspective on the history, use, and identification of synthetic food dyes. J Chem Educ 88:24–28
8. http://www.fda.gov/food/foodingredientspackaging/ucm094211.htm. Accessed 03 Febr 2012
9. Britt C, Gomaa EA, Gray JI, Booren AM (1998) Influence of cherry tissue on lipid oxidation and heterocyclic aromatic amine formation in ground beef patties. J Agric Food Chem 46(12):4891–4897
10. Nair MG, Wang H, Strasburg GM, Booren AM, Gray JI (2001) Method for inhibiting cyclooxygenase and inflammation using cherry bioflavonoids. U.S. Patent no 6,194,469 (27 Febr 2001)
11. Alonso GL, Sanchez, MA, Salinas MR, Navarro F (1997) Analysis of the color saffron. Alimentaria (Madrid, Spain) 279:115–127 (as highlighted in the January 2011 Chemical Abstracts Service "Colors of Chemistry" calendar)
12. Bunting RK (1987) The chemistry of photography. The Photoglass Press, Normal

13. Orna MV, Goodstein MP (1998) Chemistry and artists' colors, 2nd edn. Pressco, Weston, pp 333–338
14. Ronan CA (1992) John Herschel (1792–1871). Endeavour 16:178–181; John Herschel, William Henry Fox Talbot and Johann Heinrich von Mädler (1794–1874) seem to have all come up with the term "photography" within the same month
15. Herschel JFW (1840) On the chemical action of the rays of the solar spectrum on preparations of silver and other substances, both metallic and non-metallic, and on some photographic processes. Phil Trans Roy Soc Lond 130:1–59
16. Ihde AJ (1984) The development of modern chemistry. Dover Publications, Garden City, pp 51–52
17. Clark W (1946) Photography by infrared: its principles and applications. Wiley, New York, p 76
18. Keller E (1970) Images in color. Chemistry 43(December):6–10
19. Thirtle JR (1979) Inside color photography. ChemTech 9(January):25–35
20. Thirtle JR, Zwick DM (1964) Color photography. In Kirk-Othmer encyclopedia of chemical technology, vol 5. Wiley, New York, pp 812–845
21. http://electronics.howstuffworks.com/cameras-photography/digital/digital-camera1.htm. Accessed 04 Febr 2012
22. Archer RD, Cumming WG, Rennert AM (2010) Photochemistry: a SourceBook Module, v. 3.0. In: Orna MV (ed) The new ChemSource. American Chemical Society, Washington, DC, p 18
23. Haüy R-J (1822) Traité de Minéralogie, vol 1, 2nd edn. Bachelier, Paris
24. Stokes GG (1852) On the change of refrangibility of light. Phil Trans 142:463–562
25. Valeur B, Brochan J-C (2001) New trends in fluorescence spectroscopy: Applications to chemical and life sciences. Springer, Heidelberg, p 4 (alluding to Stokes' original paper, Stokes GG (1853) On the change of refrangibility of light, No. II. Phil Trans 143:385–396)
26. Valeur B (2008) From well-known to underrated applications of fluorescence, pp 21–44. In: Berberan-Santos MN (ed) Fluorescence of supermolecules, polymers and nanosystems. Springer, Heidelberg, p 22
27. Valeur B, Barberis-Santos MN (2011) A brief history of fluorescence and phosphorescence before the emergence of quantum theory. J Chem Educ 88:731–738
28. Einstein A (1917) Zur Quantentheorie der Strahlung. Phys Z 18:121–128
29. The Nobel Prize in Physics 1964. Nobelprize.org. 6 Febr 2012. http://www.nobelprize.org/nobel_prizes/physics/laureates/1964/
30. Nassau K (1983) The physics and chemistry of color. The fifteen causes of color. Wiley, New York, pp 133–138
31. Duarte FJ (2009) Tunable laser applications. CRC Press, Boca Raton, pp 227–235
32. Werrett S (2010) Fireworks: pyrotechnic arts and sciences in European history. The University of Chicago Press, Chicago, p 181
33. Chemical Abstracts Service 2010 "Colors of Chemistry" Calendar, January
34. Steinhauser G, Klapoetke TM (2008) "Green pyrotechnics": a chemist's challenge. Angewandte Chemie (International Edition) 47(18):3330–3347
35. Steinhauser G, Tarantik K (2008) Copper in pyrotechnics. J Pyrotech 27:3–13
36. Moffett M (1935) The one-woman show: monodramas. Samuel French, Inc., New York, p 109
37. Gambino M (2011) What secrets do the ancient medical texts hold? Smithsonian Magazine May. http://www.smithsonianmag.com/arts-culture/What-Secrets-Do-Ancient-Medical-Texts-Hold.html. Accessed 06 Febr 2012
38. Homan PG (2002) Pills and pill making, information sheet 7. Museum of the Royal Pharmaceutical Society, 1 Lambeth High Street, London
39. Several websites document some unique colored materials and their science. Please see, for example, http://www.chymist.com/ and http://www.teachersource.com/. Accessed 08 Febr 2012
40. Chemical Abstracts Service 2007 "Colors of Chemistry" Calendar, February

41. Berthelot M (1907) Synthesis of amethyst quartz: Researches on the color, natural or artificial, of some precious stones under radioactive influences. Compt rend 143:477–488
42. Ottaway TL et al (1994) Formation of the Muzo hydrothermal emerald deposit in Colombia. Nature 369(6481):552–554
43. Griffiths J (2008) Why does the Hope diamond glow red? Despite old rumors, chemistry—not a curse—is the key. Anal Chem 80:2295–2296
44. Chemical Abstracts Service 2009 "Colors of Chemistry" Calendar, April
45. Nassau K (1983) The physics and chemistry of color. The fifteen causes of color. Wiley, New York, pp 133–138, 174
46. Butler A, Nicholson R (1998) Hope springs eternal. Chem Br 34(12):34–36
47. http://www.si.edu/Encyclopedia_SI/nmnh/hope.htm. Accessed 13 Febr 2012
48. Shakhashiri BZ, Williams LG, Dirreen GE, Francis A (1981) "Cool-light" chemiluminescence. J Chem Educ 58:70–72
49. Sacksteder LA (1990) Photophysics in a disco: Luminescence quenching of quinine. J Chem Educ 67:1065–1067
50. Chemical Abstracts Service 2002 "Colors of Chemistry" Calendar, January
51. Zhang J, Frankel GS (1998) Paint as a corrosion sensor: acrylic coating systems. Mater Res Soc Symp Proc 503(Nondestructive Characterization of Materials in Aging Systems):15–24
52. Michalet X, Pinaud FF, Bentolila LA et al (2005) Quantum dots for live cells, in vivo imaging, and diagnostics. Science 307(5709):538–544
53. Lin C-AJ, Liedl T, Sperling RA et al (2007) Bioanalytics and biolabeling with semiconductor nanoparticles (quantum dots). J Mater Chem 17:1343–1346
54. Dyer AL, Thompson EJ, Reynolds JR (2011) Completing the palette with spray-processable polymer electrochromics. ACS Appl Mater Interfaces 3(6):1787–1795

Author Biography

Mary Virginia Orna is Professor of Chemistry at the College of New Rochelle, New Rochelle, NY (mvorna@cnr.edu), although her professional history includes service in government, not-for-profit institutions and industry. She received her B.S. in Chemistry from Chestnut Hill College and Ph.D. in Analytical Chemistry from Fordham University. She has lectured and published widely in the areas of color chemistry and archaeological chemistry and, following a sabbatical leave at the Institute of Fine Arts, New York University, she has maintained a research relationship with art historians in the area of pigment content of medieval manuscripts. Among her many professional activities, she has presented plenary lectures and named lectureships on at least a dozen different occasions. She is a tour speaker on the roster of the American Chemical Society and has been an invited lecturer to every part of the United States and many countries in Europe, the Middle East, and the South Pacific. Her many publications have appeared in the *Journal of Chemical Education, Color Research and Application, Studies in Conservation, Analytical Chemistry, Microchemical Journal, Journal of Biological Chemistry*, American Chemical Society monographs, and various other journals. She has also authored numerous book chapters and encyclopedia articles, three books and co-edited eight others. She is active in several divisions of the American Chemical Society, having served as Chair, Program Chair and Treasurer of the Division of the History of Chemistry. She is currently serving as ACS Councilor and a member of the ACS Council Policy Committee. She has received various awards among which are the Chemical Manufacturers Association's 1984 Catalyst Award for excellence in college chemistry teaching, the 1989 Council for the Advancement and Support of Education (CASE) New York State Professor of the Year, National Gold Medalist Award, the 1989 Merck Innovation Award, the 1996 James Flack Norris Award for Outstanding Achievement in the Teaching of Chemistry, the ACS 1999 George C. Pimentel Award in Chemical Education, the 2008 Henry Hill Award, and the ACS 2009 Award for Volunteer Service ACS.

M. V. Orna, *The Chemical History of Color*, SpringerBriefs in History of Chemistry,   145
DOI: 10.1007/978-3-642-32642-4, © The Author(s) 2013

During the academic year 1994–95, she was a Fulbright lecturer in Israel where she taught courses in color chemistry at Shenkar College of Engineering and Design.

Index

A

Absorption, 12, 18, 21, 23, 33, 36, 42, 43, 62, 94, 95, 96, 105, 114, 132, 136, 140, 141
Absorption band, 23, 34, 37, 41, 42, 60, 96
Absorption edge, 44
Absorption maximum, 36, 43, 44
Accum, Frederick, 129, 130
Adsorption series, 96, 97
Adulterant, 130
Afghanistan, 6, 59
Alizarin, 4, 8, 55, 56, 82, 86–89, 111, 121
Allowed transition, 135, 136
Almaden, 7
Alsace, 79
Aluminum beryllium silicate, 139
Amethyst, 138, 139
Ångstrom, Anders Jonas, 102
Aniline, 71–75, 79–81, 89, 113, 116, 118, 142
Aniline blue, 79, 80, 82, 89, 118
Aniline purple, 73
Aniline red, 79–81
Anthracene, 82, 86, 87, 89, 140
Anthranilic acid, 84
Anthraquinone, 83, 86, 87, 89
Antigen, 121
Antivitamin, 125
Archaic Mark, 57, 58
Aristotle, 12, 132
Armenian manuscript, 57
Arnaud, Albert Léon, 125
Arsenic, 9, 48, 57, 79, 116
Arsonic acid, 79
Arsphenamine, 116
Atomic absorption spectroscopy, 101
Atomic number, 30–32
Atomic structure, 30, 31, 33
Atomic theory, 29–31

Atoxyl, 116
Auxochrome, 35, 37, 38, 83
Azo compounds, 36, 82
Azo dyes, 83, 89, 115, 120
Azobenzene, 36
Azurite, 4, 5, 56, 57

B

Badische Anilin- und Soda-Fabrik (BASF), 75, 83, 84, 89, 90
Baeyer, Adolf von, 83, 84, 87, 89, 94, 112, 119
Balmer, Johann, 30, 31, 103
Band gap, 42, 43, 141
Barite, 135
Barium, 63, 137
Basel Institute for Immunology, 121
Basophilic, 113
Basov, Nikolay, 136–137
Béchamp, Pierre Jacques Antoine, 116
Béchamp's anilide, 116
Becquerel, Alexandre Edmond, 136
Becquerel, Henri, 30
Beer, August, 105, 106
Beer's Law, 105, 106
Benzene, 79, 80, 82, 83, 87, 89, 119
Berichte, 88, 96
Bernard, Claude, 112
Berthelot, Marcellin, 139
Beryl, 139
Berzelius, Jöns Jacob, 112, 122
Bethe, Hans, 39
Binder, 51, 53, 57, 63
Blue diamonds, 140
Body paint, 47
Bohr, Niels, 30
Boron, 137, 139, 140